Evolution of the Genus *Homo*

WILLIAM HOWELLS
Harvard University

ADDISON-WESLEY PUBLISHING COMPANY
Reading, Massachusetts · Menlo Park, California
London · Amsterdam · Don Mills, Ontario · Sydney

This book is in the
Addison-Wesley Modular Program in Anthropology

Editorial Board

PAUL T. BAKER
The Pennsylvania State University

JOSEPH B. CASAGRANDE
University of Illinois, Urbana-Champaign

WARD H. GOODENOUGH
University of Pennsylvania

EUGENE A. HAMMEL
University of California, Berkeley

Copyright © 1973 by Addison-Wesley Publishing Company, Inc. Philippines
copyright 1973 by Addison-Wesley Publishing Company, Inc.

All rights reserved. No part of this publication may be reproduced, stored in a
retrieval system, or transmitted, in any form or by any means, electronic, mechanical, photocopying, recording, or otherwise, without the prior written permission of the publisher.
Printed in the United States of America. Published simultaneously in Canada.
Library of Congress Catalog Card No. 73-9124.

ISBN 0-201-11633-2
    FGHIJKL-CO-79876

Contents

About the Author

William Howells received the S. B. and Ph.D. degrees from Harvard University. At present Professor of Anthropology at Harvard, he has also taught at the University of Wisconsin, Madison. His published books include *Mankind in the Making* (revised edition, 1967), *Cranial Variation in Man* (1973), and *The Pacific Islanders* (1973). Dr. Howells is especially interested in the study of human evolution, the analysis of human and population variation, and the physical anthropology of the Pacific. He is a member of the National Academy of Sciences and the American Academy of Arts and Sciences.

Genus and Species:
the Practical Meaning

"How old is man?" An anthropologist, knowing
that he is expected to give a straight-from-the-
shoulder answer, is nevertheless obliged to say:
"It depends what you mean by 'man'." To the ques-
tioner this may seem like quibbling. Man of today
is so clearly different from his nearest relatives,
the chimpanzee and gorilla, that there is no reason
to put quotation marks around him. Functionally
he walks habitually upright, speaks in a smooth
flow of vowel sounds modulated by consonants, and
makes tools, as well as tools to make other tools.
It is because so much is now becoming known about
our recent evolution that the question of what is
"human" comes up.

Living man, Homo sapiens, is a perfectly
good example of a zoological species, and a successful one. The species is polytypic, meaning
that it varies among its local populations in some
heritable features, giving rise to racial distinctions but not to an actual structure or set of races
which anyone has ever been able successfully to

enumerate or define. For these populations are all able to interbreed with complete fertility no matter how distant their homelands in recent time. This fact, and the infertility of man with any other species, are the marks of an animal species; i.e., a species manifests a set of interfertile if differing populations (or subspecies), and reproductive isolation from other species.

Homo is also a good zoological genus, though it comprises only one species today, sapiens (it is a monotypic genus, not rare in the animal kingdom). A species is the only category which is self-revealing in biological terms, because reproductive isolation is an absolute biological phenomenon. A genus, on the other hand, is a set of species which zoologists agree should be put together because of general likeness in structure, with the implication that they share some basic adaptive niche, or pattern of exploiting the environment, different from that of other genera. Man's adaptive niche, because of his cultural abilities, is enormously broad, and of course is shared by no other species. (It is in fact so distinctive that man is given a separate family, Hominidae, in addition to a separate genus.) By comparison, the adaptive niches of chimpanzee and gorilla (broadly speaking they are forest-living fruit and plant users respectively) are so similar to one another, something expressed also in their morphological closeness, that some zoologists put the two species in a single genus, Pan, instead of recognizing Gorilla as a separate genus.

To be formal about it, "species" and "genus" as defined above are terms of neozoology, i.e., applied to living animals. But the same terms,

and the same attempts to define relationships,
must be applied to animals of the past, known only
from their fossilized remains, since we cannot
comprehend such extinct animals except as forming a web continuous with animals of the present.
However, their living, functional behavior, or
their diet, or their climatic adaptation, can be in-
ferred only from carcful study; and their fertility
or infertility with other past species is untestable.
Accordingly, fossil species must be judged and
named purely from their morphological features.
This sets an obvious problem for paleontologists:
at what point, as a species gradually evolves over
time, has it become morphologically so unlike its
earlier form that it should be recognized as a new
species? The problem seldom appears acute for
paleontologists, with their long series or long time
spans. But we must realize—and this bears on
Homo—that apparent distinctions often result simply from a gap in the fossil record, separating an
earlier and later form and thus reducing confusion
only because the information is incomplete.

So much has now been discovered in the case
of man, or at least of the hominids (members of
the family Hominidae, which takes in all forms in
his special ancestral line) that both of the principal
difficulties apply. For the first, we are quite unable to say, for example, whether populations of
Neanderthal man of Europe and those of modern
man could have interbred with full fertility or not,
though the presumption is strong that they could
have. For the second, we are now finding it
harder, with gaps in the fossil record continually
being filled, to judge at what points evolutionary
changes would justify recognizing a new species or

genus. When the first really primitive human fossil was found in Java, in 1891, the separation from Homo sapiens appeared so well marked that its discoverer, Eugene Dubois, in 1894 named it Pithecanthropus erectus—new genus, new species. Because of such visible differences among the first few early men to be brought to light (including the iniquitous Piltdown forgery), and because of lack of concern over the difficulties I have described, such new forms were readily given names suggesting genus distinctions, to say nothing of mere species. Now, however, with a better framework of time, and with greater awareness of early culture and of the basic unity of the human adaptation to nature among at least the later fossil forms, there is a general tendency among scholars to recognize a single genus Homo, for "man" of the Middle and Upper Pleistocene, with two successive stages as species, erectus and sapiens.

This itself is a matter of convention, to suit knowledge of the present, rather than a final solution of classification and of lines of ancestry. Only living man, as I said, has full title to the species name Homo sapiens. Homo erectus and Homo sapiens at the moment make convenient and sensible divisions, because we have a number of clearly primitive fossil men from the earlier Middle Pleistocene separated by a fairly empty time span from the large-brained men of the later Upper Pleistocene. There are, however, some important fossils from the interval, and they are not easy to place or to interpret; and as more are found the picture will become less simple. Also, it is true that the very luck of the draw, as to which fossil was found first, which later, has necessarily led to hypothesizing and to sets of mind which would have been different if the luck itself had been different.

The point is of interest only in looking to the future. Indeed, all this introductory matter is meant to be a warning that, with fossils coming to light at an ever faster rate, present wisdom must be regarded as just that: present wisdom, not scriptural truth suitable for framing. Nevertheless, now that we know so much about dates, and have real samples of some fossil populations, it is necessary to set up a framework of interpretation, and to use the more isolated fossils as well.

One more point: we must try to apply principles from modern evolutionary theory in reasonable fashion to our hypotheses and interpretations. Such theory, for example, enforces the premise that, for any stage of development of hominids or their forerunners, there was an antecedent stage making the new stage possible structurally and behaviorally, and that both stages, and any which was transitional between them, were in successful ecological and morphological adaptation. This is a fancy way of saying we should assume the shortest evolutionary pathways to be traversed, avoiding quaint theories about tarsius-like ancestors, or brains evolving before there was anything for them to do (as was reasoned in the case of Piltdown man). Such a principle today would seem to counsel against assuming an early hominid ancestor making a drastic change from tree to ground life, finding himself at once beset by overwhelming dangerous carnivores, and surviving only by rapidly becoming a tool-user and club-wielder. Instead we must take note that man has very close

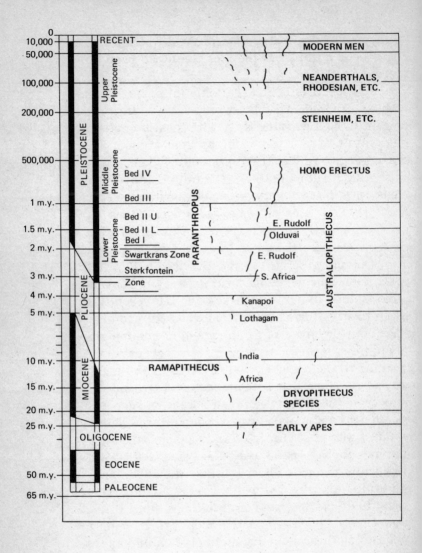

TIME CHART OF PRINCIPAL FOSSILS

This chart gives the time relations of fossil hominoids according to the age estimates used herein. The fossils or fossil groups are shown as short segments of lines at the probable time positions. These segments are intentionally not connected or lined up in succession, to emphasize that we cannot be positive about direct ancestral connections, and that we should recognize that populations shown may have come from somewhat different ancestral populations, of any general species, from any which is now known.

The time scale is logarithmic with increasing age, i.e., expanded at the recent end and increasingly compressed as one goes back through the whole Tertiary era to its beginnings in the Paleocene period. Two separate staffs, or scales, are shown at the left for the major periods. The reason is that in recent years marine paleontologists (left staff), using molluscs and fossil plankton, have disagreed with the paleontologists (right staff), who use land animals, particularly mammals as to where to place divisions—especially between the Miocene and Pliocene periods—in grouping successive faunal stages. The disagreement has little effect on the fossil hominoids, which are now positioned to a considerable extent by radiometric datings in absolute years.

living relatives in the chimpanzee and gorilla, by all signs of anatomy, tooth form, and molecular structure. We should therefore take our departure from the assumption that our first recognizable ancestors were very close indeed to theirs, that separation between the ancestral lines must not be pushed back further than the fossil record requires, and that paths of subsequent adaptation for both lines have been marked by the least dramatic episodes which we can postulate.

The First Hominids

RAMAPITHECUS

Fossils acknowledged to be clearly those of <u>Homo</u> appear something like one million years ago. This is modest compared to the age of the first probable hominid, <u>Ramapithecus</u>, dated as far back as 14 million years, in the Miocene of Africa, and coming down to about 10 million years, or even later, in specimens from India (Simons, 1969).

Today the fossil hominoids (that is, members of the common ape-human group as a whole) belonging to the later Tertiary (the Miocene and Pliocene periods) appear to have been fairly well sampled by the paleontologists. Finds come from most parts of the then tropical Old World: western Europe (Spain, France, Germany) through East Africa and into southern Russia, India, and south China. They can be grouped into four lots, and we can say that, if the group containing man's ancestors is not one of these, then that other group should also have been found by now.

9

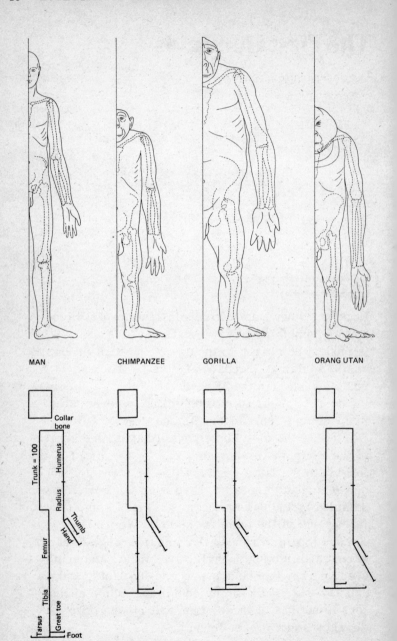

MAN   CHIMPANZEE   GORILLA   ORANG UTAN

BODY SIZE AND PROPORTIONS IN MAN AND THE GREAT APES

The upper figures are drawn to the same scale, showing man, chimpanzee, gorilla, and orang (plucked to the skin) as they would appear if average males were standing side by side. For comparison, the legs and feet are drawn in an unnatural position, and the toes of apes, especially the orang, would normally be flexed. (Adapted from a drawing by A. H. Schultz.)

The lower figures are diagrams of proportions of trunk and limbs, computed from several different studies published by Schultz. The length of the trunk (from the top of the breastbone to the top of the pubic symphysis of the pelvis) has been made the same in each case, in order to show relative length of the limbs (head, neck, and hip width have not been so scaled, and hip and shoulder joint locations are not placed in the natural position). The foot joins the leg at the base of the toes (the tarsal-metatarsal joint), not at the ankle joint.

The great length of man's legs makes his figure appear much taller than the others; the gorilla is actually as tall in absolute terms because of his long torso. Man is relatively about as long-armed as the African apes, but is shorter in the hand and longer in the thumb. The orang is especially long in the arm, fingers, and foot, with a comparatively short thumb and great toe.

Best known are the dryopithecines, originally given many different names by their discoverers but now assembled (Simons and Pilbeam, 1965) into one genus, <u>Dryopithecus</u>, including the important subgenus <u>Proconsul</u> of the Miocene of East Africa. The remains are mostly jaws and teeth only, which, however, clearly appear as the source of the dentition of man and of the chimpanzee, gorilla, and orang utan. These large living apes have become structurally modified for brachiation, habitual suspension of the body by the arms in trees, although the chimpanzee and gorilla are in fact mainly terrestrial now, supporting the body partly on the knuckles of the hand.[1] What is known of dryopithecine skeletons (mainly of the subgenus <u>Proconsul</u>) suggests a lighter and less specialized body form for these earlier apes, and a more monkey-like kind of locomotion.

A second group was that to which the modern gibbons belong, though the known fossils lack the long arms of the living animals, who are brachiators par excellence. A third group was <u>Oreopithecus</u> of Italy and East Africa, a tailless, stocky-bodied animal with arms apparently adapted for brachiation but with a short pelvis and knock-kneed legs suggesting a definite capacity for bipedal walking (see Kurth, 1967). Other manlike features were a short face, with small incisor and canine teeth, quite unlike any other ape. But the cusp

---

[1] Chimpanzees build nests and sleep in trees and find the bulk of their food there. The young of both species are more inclined to brachiation and tree-climbing in general than are the adults, though all are thoroughly adept in trees. I am quite willing to see this as indicating greater arboreality and brachiating propensities in the stock immediately ancestral to gorilla and chimpanzee.

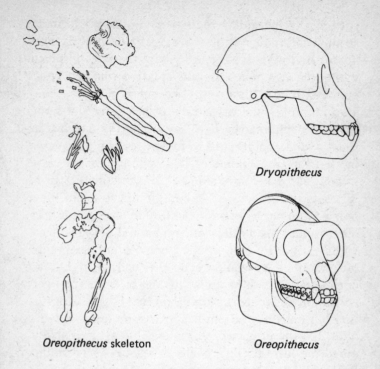

Dryopithecus

Oreopithecus skeleton   *Oreopithecus*

FOSSIL HOMINOIDS: *DRYOPITHECUS* AND *OREOPITHECUS*

Dryopithecus (Proconsul) africanus, of the Miocene of East Africa, represents a generalized ancestral pongid, but may in fact be a direct ancestor of the chimpanzee, though considerably smaller in size. The skull differs from modern apes in its lack of strongly developed brow ridges and relatively smaller incisor teeth, among other things. (After a reconstruction by P. R. Davis.)

The skull of *Oreopithecus* of Tuscany, Italy, in spite of its short face and small canine teeth, evidently had a sagittal crest along the midline. (After a recent reconstruction by F. Szalay and A. Berzi.)

In the view of an *Oreopithecus* skeleton found in 1958, the long arms suggest an adaptation to brachiation, which would be precocious compared to other known hominoids. At the same time the thigh bone (femur) suggests some adaptation to bipedalism, especially in the lateral angulation of the knee joint.

patterns of the molar teeth were entirely distinct
from those in dryopithecines, pongids (the living
apes), and man, and it is evident that Oreopithecus
was actually a more distant relative who has become extinct. His un-apelike small canines are
also striking, and the nature of his dietary adaptations is problematical. They may have had limitations which contributed to his extinction. It may
be that the importance of this peculiar fossil "ape"
is not yet appreciated: if he had the constellation
of traits described—a brachiator above and a biped
below, as usable gaits—possibly even earlier than
other lines, then study of him should be pursued
vigorously. He has significance for our own evolution, since he would argue that body uprightness,
conducive to both brachiating and bipedalism, was
an important trend in later hominoids generally.
And his convergence in body form on later hominoids like the chimpanzee makes it easier to understand the similar (but perhaps lesser) convergence
of the gibbons in the same way, although gibbon
ancestry has surely been separate from that of the
large pongids for 25 million years or more. In
spite of all this, because he was contemporaneous
with the Dryopithecus clan, obviously ancestral to
ourselves, Oreopithecus must be ruled out of any
possible place in the direct human lineage.

Ramapithecus was almost classified as a true
hominid in 1934, when his first fragment was described. But he was named simply as another
"god-ape," along with Bramapithecus, Sivapithe-
cus, etc., when such names were being handed out
to new dryopithecine finds in north India. His
recognition came in the 1960s. Louis Leakey
found a very similar upper jaw fragment at Fort

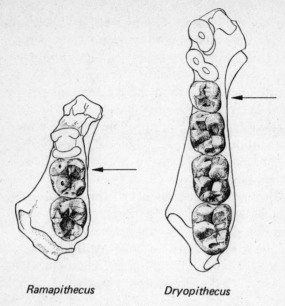

Ramapithecus   *Dryopithecus*

LOWER JAW OF *RAMAPITHECUS*

A view (from above) of part of a lower jaw from India assigned to
Ramapithecus, compared with a *Dryopithecus* fossil jaw from the
same time and place. The *Ramapithecus* molar teeth are shorter
from front to back. Especially significant is the whole jaw form.
Although both jaws are broken, the point at which the lower
border turns in (arrows) to form the chin region makes it clear
that the jaw in *Ramapithecus* was much shorter than that in
Dryopithecus. (Reproduced from E. L. Simons, "On the Mandible of *Ramapithecus*," *Proceedings of the National Academy
of Sciences,* 51 (3): 528–535, 1964, by permission.)

Ternan in Kenya. Elwyn Simons and David Pilbeam at Yale were then reducing the dryopithecine fossils to the single genus Dryopithecus by showing their high degree of similarity. It was partly this simplification which allowed the distinctions of Ramapithecus to be seen more clearly; and a number of further fragments of upper and lower jaws, from both India and Africa, previously thought to be dryopithecine, were added to those already accepted as Ramapithecus.

The distinctions are slight but important. The original upper jaw showed that Ramapithecus had a short face compared to any ape; this and the squarish molar teeth with their low cusp relief were what had looked hominid right at the beginning. It is now known that the canine tooth was relatively small, and the incisors small and narrow (Andrews, 1971), the tooth row being arched in hominid fashion. The chin region was shallow and very sloping, not "human" in today's terms, but unlike any pongid. All this, plus unusually thick enamel on the molars, and strongly implanted roots in the cheek region, together with other details of bony parts of the jaws, argues a marked side-to-side component in chewing motions and an emphasis on grinding with the back teeth, something more characteristic of hominids than of pongids (see Simons and Pilbeam, 1972). At the same time, Ramapithecus had dryopithecine traits, such as a "semisectorial" first lower premolar tooth, with a forward edge against which the upper canine bore, this being a key distinction of pongids from hominids, in whom this premolar tooth is bicuspid. A mathematical study (Bilsborough, 1971; see also Conroy, 1972) likewise fails to make Ramapithecus

particularly close to modern man, as compared
with chimpanzee and Proconsul. But this hardly
negates the probable status of Ramapithecus as a
hominid. While he displays a specific set of distinctions from contemporary dryopithecines, any
animal so close to the separation of hominids and
pongids should be expected to resemble an ancestor
of the chimpanzee more than he does a modern man.

THE BACKGROUND OF HOMINID EMERGENCE

Fossils are one thing, causes of evolution another.
Having only jaws and teeth of Ramapithecus, we
lack direct evidence of his general adaptations.
Other animals in the deposits suggest an environment of semi-open forest cut by streams. Was
Ramapithecus arboreal or not? Beyond question
man's ancestors were tree-livers, adapted to
brachiation or arm suspension to a clear degree:
this shows in many general features of his anatomy
such as the broad chest and shoulders, shortened
lumbar spine and stocky body, and various details
of shoulder and elbow. Now Ramapithecus was
either in such a stage or had already passed
through it; we do not know. We can only say that
his near relatives, the dryopithecines, were evidently less strongly specialized,[2] especially in

---

[2] There is some disagreement over the few skeletal parts, which are
largely those of *Dryopithecus* (*Proconsul*) *africanus* of the Miocene.
According to Lewis (1972) this form exhibited special features of
the wrist shared only by the living large apes and man, suggesting to
some (Conroy and Fleagle, 1972) that the pattern of brachiation and
knucklewalking was present incipiently. Preuschoft (1973), analyzing limb bones biomechanically, finds strong indications that *Proconsul* was essentially quadrupedal, like monkeys, either in trees or

limb proportions, for brachiation than were later apes, and the same may have been true of Rama-pithecus.

The primary shift, the first hominid adaptation, is likely to have been dietary, always an obvious bet for an evolutionary departure. And this is where we have concrete evidence in teeth. The living apes, especially the chimpanzee, in recent evolution have emphasized the teeth at the front of the muzzle—incisors and canines (Conroy, 1972). This is evidently an adaptation to diets of tough fruits or, for the gorilla, of coarse vegetable stems, edible bark, and leaves. This plan of dentition, less emphasized, was present in the dryopithecines, who were probably more general herbivorous feeders. Ramapithecus apparently took another direction, of diminished fore teeth and heavier molar grinding. Jolly (1970) has suggested that this and associated features are related to a "graminivorous" diet, of small, tough, but nutritious objects like seeds and roots. He bases this on a comparison of tooth, jaw, and skull differences between the common savannah baboons and the gelada baboon. The latter is more highly terrestrial, and more closely specialized than the others in a diet of grass blades and rhizomes, the bulbous stem parts of grasslike plants. This diet needs less work by the incisors and more by the molars. In the gelada the molars are relatively larger and the incisors and the canines relatively smaller. Early hominids, he suggests, had a

---

on the ground. This would argue that brachiation developed in a common ape-human ancestor during the five or six million years between *Proconsul* and *Ramapithecus* to the point of leaving a much more definite stamp on the skeleton.

parallel adaptation, but one more toward seeds or
grains than rhizomes, reflected in broad molars
worn flat, without the surface complexity of gelada
molars.

Such a diet, which need hardly have been an
absolute specialization, points to the ground rather
than the trees. It would fit a natural ecological
displacement, in a gradually deforesting environment, which could have caused Ramapithecus to
diverge from the dryopithecines and from the less
specialized, more arboreal, herbivorousness of
the latter. The rest of the interplay between the
early hominids and ground life is now obscure for
lack of evidence, though we should remember not
to see it in terms of the present. For example, it
is doubtful that carnivorous predators were then
the threats they would be now, since the great cats
had not evolved and the greatest danger might have
come from ancestral hyenas or a huge bearlike dog
(Kurten, 1972). So we need not be perplexed at the
apparent defenselessness of a small groundling.
This applies especially to the reduced canine of
hominids, always a stumbling block to theorists.
In male apes (and dryopithecines) a large canine
is supposed to serve as a weapon of defense or at
least of threat; but it is also probably useful for
stripping coarse vegetable food as well, particularly to the gorilla. If, however, in hominids the
canine tooth had a positive selective reason for
shortening--accommodating itself to greater
lateral motion in chewing--the paradox is removed.

The Australopithecines

MORE ABOUT CHEWING

If there was any progressive evolution in <u>Rama-</u>
<u>pithecus</u> between about 14 and 10 million years ago,
we do not know it; and we know nothing at all after
that until about five million years ago, when the
first signs of the australopiths appear. Following
this the record is virtually continuous to the present. Remains of the australopiths are now copious
and steadily increasing. It is ironic that the whole
first stage of recovery was in South African lime-
cemented cave fills, where the fossils are very
difficult to extract. Much later the search began
for East African specimens eroding naturally out
of the sediments of Olduvai Gorge, or of the Omo
River and other tributaries of Lake Rudolf in Kenya
and Ethiopia. Not only are these specimens often
in beautiful shape, but they usually lie in a stratigraphy—bedded history—which gives their time
relations, and sometimes also contains volcanic
material which can be absolutely dated by the
potassium-argon method.

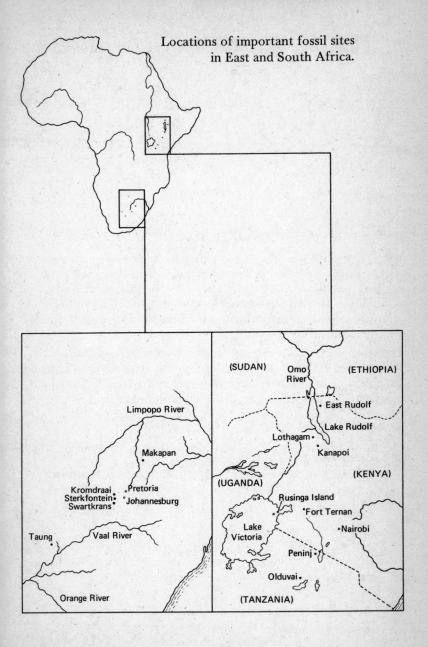

Locations of important fossil sites
in East and South Africa.

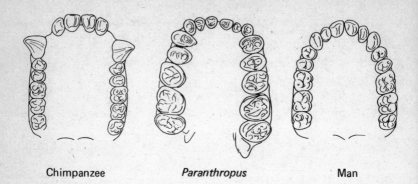

Chimpanzee   *Paranthropus*   Man

UPPER JAWS OF CHIMPANZEE, *PARANTHROPUS*, AND MAN

The upper tooth row in the chimpanzee is shaped like a croquet wicket, not like an arch as in modern man. This shape is due to the long, parallel rows of cheek teeth, the large canine teeth at the angles (with a gap or diastema just in front of them to receive the lower canine tooth), and the broad incisor teeth forming a straight line across the front.

The upper teeth of *Paranthropus* (drawn from "Zinjanthropus," or OH 5, from Olduvai) contrast profoundly with those of pongids, and might be said to have exaggerated "human" traits. The molar teeth, though very large, emphasize breadth rather than length, while the canines are small and nonprojecting, with no diastema between them and the relatively very small and narrow incisors.

The tooth row of *Australopithecus* (not shown) is somewhat more like that of *Homo* in its characters and proportions.

The commonest parts, again jaws and teeth, are resoundingly hominid, exhibiting traits specified by Simons for Ramapithecus and by Jolly for the graminivorous adaptation (see Conroy, 1972). In fact there seems already to have occurred an adaptational split, into Paranthropus, with exaggerations of these hominid traits going beyond modern man, and Australopithecus, more generalized as to teeth and more like Homo in this and other ways. The split had happened certainly before three million years ago, and much more probably before five million years ago. Let us make preliminary comparisons between them, based mainly on fossils found in South Africa, where those of the Australopithecus line are less advanced evolutionarily and are probably early in time.

Paranthropus had jaws heavier than a gorilla's, but deeper and short and showing a totally opposite mechanical orientation: toward the back, not the front. Incisors and canines are small and narrow, the incisors in some cases showing crowding like that often seen in modern man. Behind the canine is a radical shift: the molars are very large, and the premolars as well have the great breadth of the molars, with the whole providing an impressively large milling area marked by flat wear. Suppose you express the area of a tooth surface by simply multiplying length and breadth, and show the area of the lower canine as a percentage of the first premolar just behind it. These are comparative figures[3]:

[3] Based on figures from Wolpoff (1971). Using figures from Robinson (1956) for smaller but more specific groups of *Australopithecus* and *Paranthropus* respectively from the sites of Sterkfontein and Swartkrans in South Africa only, gives 93% and 55%, a slightly greater contrast between the two forms.

U. S. whites	104%	(canine slightly larger)
Homo erectus	89%	
Australopithecus	83%	
Paranthropus	53%	

The distinction between Paranthropus and Australopithecus, as well as later men, is evident.

This milling surface in Paranthropus is set in a deep, wide jaw slung under a high face. Some larger individuals have a gorilla-like development on the skull: a bony crest formed along the midline of the top of the skull serving as increased bone surface on the small cranial vault for the origin of the large temporal muscles to the jaw. This crest, especially, once led some anthropologists to suggest a close affiliation with apes, but the appearance is deceptive. In a gorilla there is an emphasis on heavy projecting canine teeth and broad incisors, useful in stripping vegetable matter. The fibers of the temporal muscle which do most of the work at the forepart of the jaw are those running from the top of the mandible farthest back on the side of the skull, so that they are opposite their point of work; and so the midline crest in a gorilla rises to its greatest height here, abutting the great plate of bone which forms the back of the skull and to which are attached the neck muscles holding up the head from the rear. In australopiths, on the contrary, it is the back teeth that are heavy duty chewers, the fore teeth being nibblers, however powerful. On the skull's top the bony crest, when present, served the temporal muscle fibers for the back of the tooth row, pulling vertically from the fore part of the skull; and so

the crest is set well forward, disappearing at the
back. It makes no connection with the line of attachment of the neck muscles, since this line is
far down the skull at the rear, reflecting a head
held erectly on the neck and requiring much less
muscle to keep it in balance.

The South African crania assignable to Australopithecus are fewer and more fragmentary. They
are higher and rounder in the vault, and probably
seldom had a central (sagittal) crest. Molar teeth
were of impressive size, though not as large as in
Paranthropus, and their contrast with the front
teeth was less. The conclusion of many is that
Paranthropus was the more specialized feeder,
herbivorous and, following Jolly's hypothesis, a
chewer of seeds and hard roots; and that Australopithecus had remained more general in diet and
was partly carnivorous. This conforms to the
principle that two species cannot exist simultaneously in the same territory (as these two did, at
least in East Africa) unless they avoid competition
by occupying different ecological niches, or, put
simply, keeping out of each other's way in the food
department. The two species must have diverged
from a common ancestor in that same period, or
the later part of it, which saw specialization taking
place in the great apes. Thus it was a period in
which hominid evolution could still be expressed in
speciation and dietary specialization, rather than
by broadly progressive changes in such things as
manipulative ability, and by brain growth and complexity. These last things were the marks of later
human evolution, and they are the kind of change
which prevents rather than encourages specialization and speciation.

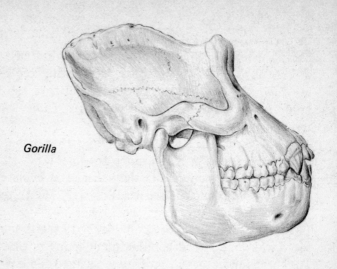

Gorilla

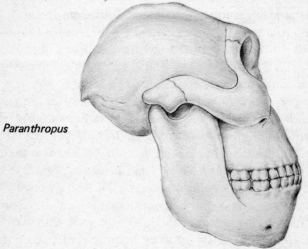

Paranthropus

SKULLS OF *GORILLA* AND *PARANTHROPUS*

The contrasting shapes are due to differences in posture of the head and in emphasis in mastication (front teeth versus back teeth).

The gorilla has a massive plate of bone at the back of the skull, rising well above the level of the braincase, for the insertion of neck muscles to hold up the head from the rear. In *Paranthropus,* the upper limit of neck muscle insertion is low on the back, and the area is small, reflecting an upright posture and a better balanced head calling for far less muscular power in the neck.

In *Paranthropus* the face is long but flattish, and the jaw is massive and deep but short. The power of the lower jaw is exerted mainly on the molar teeth. Power is partly supplied by the temporal muscle, which is attached to the region of the coronoid process (the forward part of the ascending branch of the mandible), and which spreads fanwise over the side of the skull. Because of the large size of the muscle in these two animals, an extra crest of bone is developed along the midline to give it the necessary area of attachment. In the gorilla this sagittal crest is highest at the back (where parts of the temporal muscle work to power the front teeth), merging with the nuchal plate for the neck muscles. In *Paranthropus* the crest is further forward, where the temporal muscle pulls on the back teeth, and it makes no connection with the crest for neck muscles in the rear.

MORE ON GENUS AND SPECIES

Here is a viewpoint that needs emphasizing. It is
easy to get snarled in terminology and to find one's
intended biological meaning obscured by a façade
of names like Paranthropus, Australopithecus, and
Homo, basic though such naming is in establishing
order. Confusion is the easier for readers who
may not detect that different writers are applying
the names quite differently, which is the drawback.
Like most students I believe that a speciation had
occurred in the hominids in the interval between
Ramapithecus and the australopiths, and that over
the period we are now considering there existed
two separate population series. Some name them
as two species (of Australopithecus, the senior
genus); a few, as I do here, prefer two genera,
because of their probable ecological divergence
and its consequence in relative evolutionary progress. In neither case should this mean, for example, that all Paranthropus fossils must be
closely similar to all others. Given the known
time and space, considerable population or subspecies differences should be expected within each
complex, to the point of creating difficulties of
definition and description (in taxonomic terms),
especially in the case of the more rapidly evolving
Australopithecus. During this same time period
(the late Pliocene and early Pleistocene) other
mammals of the whole region were rapidly evolving and speciating, such as the elephants, the pigs,
and the baboons. So, in spite of the dogma stated
above, that "man," with the flexibility of adaptation given him by culture, has passed the point of
speciating by splitting, there is no reason to

suppose that the australopiths were not still showing such tendencies. The following discussion of
differences between East and South Africa, and of
problems of Homo versus Australopithecus as seen
in the fossils, should be read with all this in mind.

EVIDENCE OF BIPEDALISM

Both australopith species were erect bipeds. This
was first deduced from the skulls, especially the
positioning of neck muscles and of the foramen magnum (where the skull rests on the spine). The de-
duction was then fully confirmed by bones of the
pelvis and the foot, even by a lonesome last joint
of the big toe. This toe bone shows the slight
asymmetry of its joint surface, which advanta-
geously meets the stress of forces in walking and
is found only in the human foot (Day and Napier,
1966).[4] The various remains manifest an advanced
stage, if not a complete one, toward the bipedalism
of modern man.
     Fully developed human bipedalism makes a
number of demands on the shape of the skeleton.
One is an arched foot, having a great toe lined up
with the rest, not diverging as in other primates.
A good foot skeleton from Olduvai Gorge, the isolated toe bone, and another foot bone from South
Africa, all show or indicate this development. A
second feature is a leg that can readily be
straightened under the body, swinging easily backward and forward near the midline of the body,
where the center of gravity is. A third is the

---

[4] Oxnard (1971) finds that it was by no means completely modern
in form.

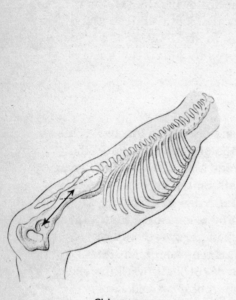
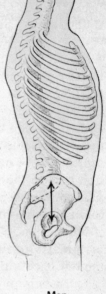

Chimpanzee   Man

THE PELVIS IN CHIMPANZEE AND MAN

This figure shows the normal position of the pelvis and torso in
chimpanzee (in the terrestrial quadrupedal stance) and man (upright). Of particular importance is the shortening of the pelvis
in its upper part, between the sacro-iliac attachment of the spine
above, and the hip joint below. The shortening in man has made
balance in bipedal walking easier, not only by lowering the center
of gravity but also by allowing different arrangements of muscles
maintaining balance. (After drawings by A. II. Schultz.)

ability to swing the leg far back with power (fully extending the hip joint); and signs of this are seen in the femur, of which several fragments show a groove at the back of the neck where, in man only, the tendon of a muscle (obturator externus) crosses it when the hip is fully extended (Day, 1969).

The pelvis reflects a whole complex of adjustments, both to the function of the legs and to the human posture of the trunk, in which the center of gravity is low and well back, slightly behind the axis of the hip sockets, instead of well forward and high, as in apes. In man the whole pelvis is shortened above (for better balance) and below, in the ischium (what you sit on), since raising of this part relative to the femur allows a fuller and more powerful extension of the leg by the muscles across the hip joint.

Other features control balance of the trunk in walking and standing (see Napier, 1967). A strong ligament (ilio-femoral) in front resists falling back. Bending back of the ilium, or upper rim of the pelvis, positions the gluteus maximus muscle of the buttock so as to resist falling forward. Semicircular lateral extensions of the pelvic rim allow other muscles (especially the gluteus medius) to act like a clamp on the side between leg and hip, to keep the trunk from falling toward the other side of the body when that leg alone is supporting it, as in striding.

If we take the African apes as an approximate ancestral model, the above is a list of differences found in man, though it is only a bare outline of matters now coming under detailed and intensive investigation in various ways: biophysical, electromyographic, etc. We may characterize the

australopiths as follows. In main features, such
as the shortened and ringlike pelvic girdle as opposed to the high elliptical pelvis of apes, they
were fully hominid, as in the basic features of leg
and foot. At the same time they lacked various
refinements of Homo sapiens, having a less shortened ischium, less rounding of the lateral extension of the ilium, differences in the upper end of
the femur, etc. The meaning of all this is not
clear, and experts disagree. Some hold that the
australopiths could run bipedally, balancing being
easier, but could not walk long distances using a
striding gait like modern man. However, with the
remains often fragmentary, and in some cases
having features seen in neither men nor apes, interpretation is incomplete to say the least. We
know only that the australopiths were unequivocal
hominids in skeleton as in teeth.

There is another source of difficulty. Which
bones belong to which form? In South Africa sites
contain skulls and teeth that have been assigned to
only one or the other: Australopithecus at Makapan
and Sterkfontein, Paranthropus at Swartkrans and
Kromdraai; and so odd skeletal parts may be assigned the same way with some confidence. But at
Olduvai the two forms were present together, and
assignment is unsure, a guess based on size.
Around Lake Rudolf, individual bones are picked up
in the open, with no evidence as to the species involved. Altogether, the idea of "robust" and
"gracile," based on skulls, may be overdone: certainly some of the "gracile" forms were very
small-bodied; but in South Africa, part of an arm
bone supposedly from a "gracile" site (Sterkfontein) may be larger than the same part from a

"robust" site (Swartkrans). As far as bones can
be correctly placed, some others think Paranthropus was distinctly less evolved toward full
bipedalism (Napier, 1964), others that there is
little to choose (McHenry, 1972). Everything suggests, however, that Australopithecus eventually
progressed beyond any stage reached by Paranthropus (see below).

ORIGINS OF BIPEDALISM

Bipedalism is a primary hominid adaptation, perhaps the basic one. Even if the first isolation from
dryopithecine pongids stemmed from a dietary
adaptation, it is clear that bipedalism was a prerequisite for bimanualism and all that flowed from
it. We have the beginnings of evidence as to the
route, to be considered next, but we can only speculate as to the reasons. The scant fossil record
suggests that in evolution the pongids have gone
from a lighter-bodied dryopithecine form, more
generalized to locomotor potentialities, to a
heavier-bodied one more specialized (in Africa)
both for brachiation and for quadrupedalism. A
gibbon, the smallest ape (though no close relative
of ours), runs erect on the ground when he has to,
keeping his arms up, instead of using the arms for
support like chimpanzees and gorillas. Now, by
and large, bipedalism is calculated by biologists
to be more expensive in energy, for distance trav-
eled, than is quadrupedalism, arguing that animal
bipeds are rare, and man exceptional, for this
reason. Surprisingly, however, young chimpanzees, weighing about 40 pounds only, trained to run

both ways on a treadmill (Taylor and Rowntree, 1973), were found to use no more oxygen as bipeds than as quadrupeds, at any speed. So a small early hominid fresh from the ancestral dryopithecine source, partly adapted by a brachiation which would encourage uprightness, but not so top-heavy as the African apes, might have embarked on a mixed tree and ground life, with both quadrupedalism and bipedalism as possible gaits.

This is the essential condition. From such a point, various reasons for adoption of bipedalism have been put forth (see Sigmon, 1971). Uprightness for speed in running is conceivable if not persuasive: man runs slowly compared to typical quadrupeds, and it is difficult to think of species of wild mammal he could possibly catch in this way (unless you want to apprehend skunks and porcupines). A gorilla on all fours can readily outrun a man, especially on rough terrain; however, gorillas have probably become adapted for this gradually, while an early hominid might originally have been able to run faster on two limbs than four. Another suggested reason is constant standing erect, in savanna grass, to spy predators, in the case of an animal equipped, like other primates, with better vision than smell.[5] Running while using the arms to carry food to a safe eating place is a further suggestion (Hewes, 1961), especially since such carrying became important in later development (see below). (Let us pass quickly over the recently revived idea that this hominid constantly sought refuge in the edge of the sea, and had to

---

[5] Kortlandt and van Zon (1969) have found that savanna-living wild chimpanzees walk bipedally more often than forest livers.

stand erect or drown.) The most reasonable assumption is this: the hominid was indeed sufficiently adapted to uprightness, behaviorally and
structurally speaking, so that frequent bipedalism
was a major option, and perhaps a less costly one
than it eventually proved for gorillas and chimps.
Quite likely no single cause was responsible alone;
rather an original small bias toward an erect gait
entailed an accumulation of adaptive rewards until
bipedalism reached the point of no return. But hypotheses as to the nature of the rewards, or pri-
orities among them, can hardly be judged from
present evidence.

EVOLUTION OF BIPEDALISM

What about the route?[6] Was it a simple transition
from chimpanzee to human bipedalism? Qualified
anthropologists use just such phraseology, though
perhaps as a manner of speaking. Or has the
chimp also shown some divergence, however slight,
from a common ancestor? The Miocene dryopithecine Proconsul apparently had a basically pongid-
like talus (the weight-bearing foot bone at the
ankle—see Day and Wood, 1969; Clark and Leakey,
1951), but with some primitive, even monkey-like
features, as well as some nonpongid traits suggesting that Proconsul could have balanced more easily
in standing erect than can a chimpanzee, and that

---

[6] What follows is a cursive and generally hypothetical rendering
based on a considerable amount of special and laborious research
over the last 40 years into many aspects of bone form, muscle arrangement, and animals in action. This has been done with much in-
sight and ingenuity, and points to considerable areas of ignorance
as to detail which further research can elucidate.

his feet rolled onto the outside edge less easily.
There is much to be learned, but this would not
contradict the idea that the earliest hominid foot
(like earlier pongid feet) might have differed significantly from that in modern chimpanzees.
Oxnard (1972), using some data of Day and Wood,
in fact finds strong signs of distinction between the
Proconsul talus and those of both apes and man.
And Preuschoft (1973), analyzing the same bones,
doubts that Proconsul used the sole of his foot flat
upon the ground like man or chimp, but instead
raised the heel slightly, like a monkey.

In various ways, too, chimps may be uncertain heralds of bipedalism in man. Though decidedly capable of upright walking, chimpanzees
and gorillas are not bipeds.[7] On the ground they
are "quadrupeds," we saw, supporting themselves
in a specific way on the middle knuckles of the
hands. (The gibbon, we also saw, does not do
this. As to the orang, the less said the better:
this highly arboreal animal, seldom out of trees,
normally uses his fists when on the ground, not
his knuckles; and his long, permanently curved
grasping side toes and very short "great" toe make
the application of his foot to the ground a problem,
generally done by resting it on the outside. He
shows none of the smartness of the terrestrial gait
of the two African pongids.)

---

[7] Standing and walking is done quite easily, and even habitually by
captive animals. Observers, including myself, have seen zoo gorillas
stroll upright, with an armful of food, like paunchy clubmen, or
with the pride of Casey at the bat, evidently as the result of training.
It is the more impressive in a gorilla because it is self-training; nobody trains gorillas. (Recent TV humor: Q. Where does an 800-
pound gorilla sleep? A. Anywhere he wants.)

Washburn (1968) believes there was a knuckle-walking stage in the hominid apprenticeship preceding bipedalism, so that the present chimp remains a proper model. Tuttle (1969) does not agree; from extensive dissections he finds anatomical specializations in the bones and tendons of knuckle-walkers missing from man. Also, from a theoretical point of view, the center of body gravity is involved. With its forward placement, and the long upper part of the ilium of the pelvis, the apes are top-heavy; and in twisting themselves erect they bend their knees in order to keep support under the weight, not because the knee will not straighten. With all this, it seems to me that a trend to knuckle-walking as a typical gait in the earliest hominids would have tended to become an adaptive trap, difficult to escape from into habitual uprightness. Therefore, though some knuckle-walking may have served such creatures opportunistically, like bipedalism in apes, it was probably not a real interlude in human evolution.

A chimp walking upright rocks his pelvis from side to side, toward the supporting leg, and swings it forward on the side of the moving leg, also swinging this leg out and well forward into position for the next step; his arms help to balance and to compensate for the rotation by swinging rather widely (Jenkins, 1972; Elftman, 1944). These pelvic gymnastics are greatly diminished in knuckle-walking, the chimp's stable adaptation. They are also minor in upright-walking man, using his stable adaptation. His pelvis is actually a little lower on the side of the moving leg. His arms swing to counterweight his legs, but easily fore and aft, with none of the flailing of a chimp, and his legs also swing directly fore and aft. (As a piece of homework, the next time you have an opportunity, look down, out of a window some floors up, at people walking on the sidewalk directly below you, to see how the body keeps its frontality easily while arms and legs swing back and forward in direct lines.) As one leg begins its stride the knee straightens (while the other bends to clear the ground with the foot); the leg then lifts the body on the ball of the foot and begins the forward thrust. That is the meaning of having a solid arched instep. This leg is performing a tiny pole vault. At the top of the step the body is raised to its highest level, and begins a downward glide until the other leg takes over: a free ride which recoups some of the energy spent in lift and push. We have seen what arrangements in the pelvis hold it generally level and erect during the performance. (More homework: observe a pair of people walking together but not in step, and note the rise and fall of the body in the contrasting bobble of their heads.) A chimpanzee moves up and down also, but he never gets his center of gravity ahead of his stride, to glide downward; it is as though he were constantly climbing out of a depression.

What about feet, also involved in this smooth legwork? Our foot is now so different from an ape's that the transition is hard to see, though no one doubts that the human foot came from a pongid grasping foot. One important aspect is that human feet meet the ground near the body's midline, another part of the center-of-gravity, smooth-striding business, since in apes the feet are planted far apart. (More homework: walk unobtrusively behind members of our species, and

observe how, in some, the inner knobs of the an-
kles barely miss one another as they pass.)

Walking either erect or on all fours, a chimp
(who has no ball to his foot) applies pressure to
the ground from his trailing foot with the great toe,
which is opposable and somewhat separated, and
with the side toes, which may or may not be curled
under (Elftman and Manter, 1935a). (It is my impression that chimps and gorillas standing or running erect separate the great toe and the others,
enlarging the spread, more than when walking, but
I do not know that this has been carefully recorded.)
The tendency of the foot to invert (the soles to face
each other, as in climbing) is strong. This leads
to pressure on the outside of the foot, to pigeon-
toeing, and to counteracting this by turning the leg
out, giving rise in walking to greater pressure on
the outside toes.

Now if, in the early erect-walking hominids,
the feet were the first element of the leg to approach the midline, resulting in a highly bow-
legged walk, the process would have given advantage to a shortening and reverse twisting of the
side toes—an essential preparation for the appearance of arch and ball of foot—while the "grasping"
arch form of the instep provided more of the basis
for the arch. Pressure from the side instead of
from above would cause less crushing of the exist-
ing arch, and less need for the great toe to provide
stability by abducting—standing off. Relieved of
constant weight-bearing, it is likely that this toe
would have had new tendencies: to draw alongside
the others to stiffen the foot, to lose its opposabil-
ity, to undergo reverse twisting of its end, and so
to join the arch on the inside of the foot. It was

perhaps during this phase that the foot was further
stabilized against both weight and the old tendency
to roll outward, by a raising of the inner border of
the talus and a leveling of its support by the heel
bone below, two features considered important in
the foot's evolution (Morton, 1935; Elftman and
Manter, 1935b).

Finally, the knees also approached one another, until it was the thigh bones that sloped inward, not the lower legs, which now became vertical.[8] The arch of the foot having become stable
enough to support weight from directly above while
functioning, the great toe became dominant in
weight-bearing and thrusting, and the foot adopted
a fore-and-aft axis which is turned slightly outward from the midline, as in modern man, rather
than inward as in apes, in order to give advantage
to this strong inner side.[9]

The above is a highly hypothetical reconstruction. It demands a small-bodied ancestor like
Ramapithecus, whose early bias in posture on the
ground ran opposite to that of the closely related
pongids, initiating a bipedalism which became set
through a shortened pelvis and a lower center of
gravity. Stabilizing the pelvis on the legs permitted the feet to use a smaller total area of
ground for balancing, letting them function more

---

[8] Although details of his foot are not known, *Oreopithecus* seems
to have had a femur of similar form, contrasting with pongids and
known fossil apes.

[9] In the tibia, the axis of the lower (ankle) joint surface is turned
inward, relative to the axis of the upper (knee) joint surface, about
9° to 10° on the average in gorillas or chimpanzees. It is turned outward in man about 18° to 20°, of course with much variation (Martin, 1928, p. 1165).

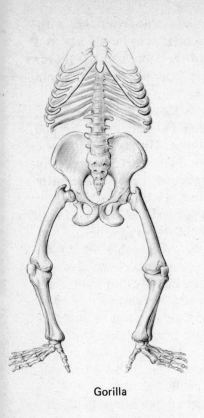
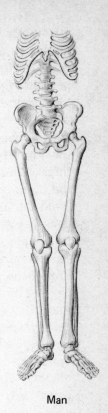

Gorilla   Man

LOWER LIMBS IN GORILLA AND MAN

This figure shows adaptations in the human skeleton for bipedal walking. The pelvis is both shorter and more bowl-like, allowing muscles at the front, side, and rear greater efficiency in moving the leg on the pelvis and stabilizing the pelvis on the leg.

The posture of the legs has changed in man to bring the knees and feet close together. In the foot, the opposed, or abducted, great toe is not used to support the inner side of the foot, which has a stable transverse arch in man, becoming instead a powerful propulsive member.

efficiently near the midline. This position, with
the knees and upper legs still separated, allowed
compacting of the arch of the foot, and this in turn
permitted weight-bearing from above, i.e., bringing together of the knees.

There are some signs that Australopithecus
had reached this last stage but that Paranthropus,
though fully bipedal, had not. In the latter, specimens from Swartkrans in South Africa suggest that
the thigh bone (femur) descended more vertically
from the hip joint (Jenkins, 1972); and the long
neck between the joint head and the shaft of the
femur also suggests special leverage for holding
the pelvis steady laterally over such a leg (by action of the gluteus medius and other muscles—see
page 31). And a lower leg bone (tibia) from East
Rudolf, which may belong to Paranthropus, has a
transverse slant across its knee joint of about 10°,
looking as though it sloped inward from knee to
ankle (although this figure is not impossible for a
modern human tibia, and the fossil bone has not
been analyzed fully). Finally, a talus from Kromdraai (a Paranthropus site) departs distinctly from
modern tali toward those of pongids in important
features (Day and Wood, 1968). Another talus from
Olduvai, part of a small but clearly "human" foot,
with a good arch, also departs from modern man
in the analysis cited, but not so much. (This foot
is supposed to belong to Australopithecus or "Homo
habilis.")

Paranthropus also had a longer, more pongid-
like ischium of the pelvis than modern man or
Australopithecus. In addition, the latter, judging
by parts of femora from South Africa, had apparently achieved a knock-kneed stance generally like

modern man. There are still gross gaps in the information, quite apart from our not knowing anything about Ramapithecus, or which bones are
whose in East Africa. For example, there are hip
(innominate) bones, and hip ends of the femur from
Swartkrans, but not knee ends or, except for the
Kromdraai talus, anything further down. Nevertheless the feeling is getting stronger (e.g.,
Napier, 1964; Robinson, Freedman, and Sigmon,
1972) that Paranthropus had definitely not evolved
as far along the trend to bipedalism as Australopithecus and then Homo, and may have remained
something of a waddler, his pelvis suggesting that
he was relatively better at powerful climbing and
less good at effortless striding.

HANDS, TOOLS, AND BRAINS

One eventual reward of bipedalism was release of
the arms and hands from any duty in locomotion,
terrestrial or arboreal (though tree-climbing is
certainly not a lost art, being almost a compulsion
in juvenile hominids today). Although it was the
foundation of all man's technical achievements, in
the australopithecines our own manual dexterity,
and the fine prehension allowed by the full opposition of the flesh pad at the end of the thumb to
those on the other fingers, were almost certainly
not yet present. Fairly good hand bones from
Olduvai suggest general human proportions. However, a metacarpal (the basal bone) of the thumb,
also from Olduvai, is interpreted by Napier (1962)
as capable of a good "power grip" but not a "precision grip." A mathematical multivariate analysis (Rightmire, 1972) of the same bone shows that

EARLY STONE TOOLS

Left: Flaked pebble from Koobi Fora area, East Rudolf, Kenya. Age about 2.6 million years.

Right: More developed tool from Swartkrans, Transvaal, South Africa.

it is far from modern in its shape, and probably had some of the functional limitations of a chimpanzee thumb (which is not saying that its owner was arboreal). If the thumb belonged to Paranthropus, as is likely, then Australopithecus (or "Homo habilis"--see below) could already have had a more advanced hand, that is, by two million B.C.

  Known bones of the arm also appear clearly hominid in some ways, particularly the four-million-year-old elbow joint of a humerus from Kanapoi near Lake Rudolf (Patterson and Howells, 1967). Other bones from the East Rudolf and Omo areas--a good humerus and an ulna, not from the same individual--are of general hominid size and

proportions, though clearly expressive of great
muscular power. A multivariate study of the humerus (McHenry, 1973) places it somewhat, but
not unequivocally, closer to those used in suspension (e.g., an orang's) than to modern human
bones. (Note once more that these bones are not
certainly assigned as to species, and it is only an
assumption that they belong to Paranthropus,
whose cranial remains are commoner among discoveries in the region.)

One of the most important recent revelations
is the finding, by Kay Behrensmeyer and Richard
Leakey (M. D. Leakey, 1970; Issac, Leakey, and
Behrensmeyer, 1971), of stone tools 2.6 million
years old in the Koobi Fora area of East Rudolf.
They are small, and made on pebbles; they must
have been used for cutting and scraping, not as
weapons. Stone tools had previously been found at
both Sterkfontein and Swartkrans in South Africa,
and in the lowest levels at Olduvai (1.8 to 2.0
million years).[10]   But this is a new and unequivocal early date. The South African sites had never
been placed in absolute chronology, and the presence of tools had been felt a paradox: the mark of
man, tools, being made by prehuman hands and
brains, supposedly incapable of making them. But,
as Washburn has long emphasized (1960), this is
another case where evolutionary theory suggests
the more natural sequence: the making and use of
tools, even by primitive hands lacking a precision

---

[10] Tools with a date of approximately two million years have now
been reported from sites further north, in the Omo Valley of
Ethiopia (Merrick et al., 1973). They differ from those found previously, being predominantly small angular fragments made of quartz-
ite, with some flake tools.

grip, provided just the selective environment for the evolutionary development of anatomically more dextrous hands.

Here we must take note of the fact that the stone tools are clearly recognizable, culturally patterned implements, and cannot represent anything like the first tools used or made. Jane Goodall has shown that chimpanzees make "tools" by trimming grass stems with which to fish termites out of their hills. And Kortlandt (see Kortlandt and van Zon, 1969) found that savanna-dwelling chimpanzees readily used "clubs" of one sort or another to attack real or dummy leopards, in zoos or in the wild, which forest chimpanzees did not do. Dart has for years (e.g., 1949) proposed that many antelope humeri and mandibular parts found at Makapan in South Africa represent, respectively, clubs and knives or scrapers employed by Australopithecus. Whether or not his interpretation is correct in detail—and many have disputed him—it is quite logical that such pre-shaped objects, modified by a few particular ways of breaking them to form new shapes, should have been a step on the road leading from the use of randomly shaped materials (e.g., broken stones) to the patterned shaping of tools from things like pebbles, which are mere raw material and do not suggest a tool form or use in their raw shapes.

Another element in the apparent paradox I mentioned is australopith brain size. For the South African specimens of Australopithecus, estimates of internal skull capacity based on the most exact reconstructions (Tobias, 1971a; Holloway, 1972) average less than 450 cubic centimeters. These are chimpanzee levels. Sizes for

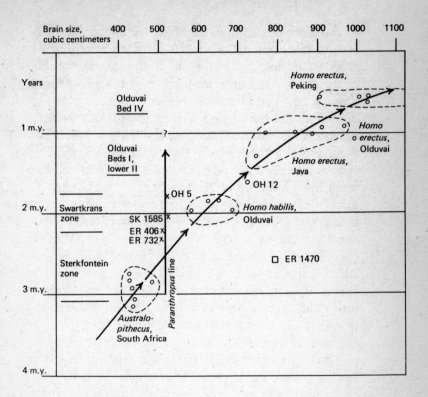

BRAIN SIZE AND TIME

This chart illustrates in a general way the increase in brain size from *Australopithecus* of South Africa, through *"Homo habilis"* to *Homo erectus*. Individual fossils in this line are shown as small circles. Individuals of *Paranthropus* are shown by "x," and the vertical arrow relates to them, suggesting the absence of any clear trend of increasing brain size in this lineage. The anomalous position of ER 1470 is obvious.

In most cases both date and brain size are based on estimates only, however good. The chart is therefore not as exact as it may appear. Estimates of brain size are largely those made by Holloway and Tobias in various writings.

Paranthropus so far estimated, including East African finds, are all near 520 cc, also not impressive. This had led to the suspicion that the tools in the South African sites indicated the presence of a more advanced hominid, namely Homo, who, however, did not appear among the fossils except possibly, as we shall see, in the case of "Telanthropus" at Swartkrans. But, considering their age, the early East African tools cannot have been made by a hominid of Homo erectus level. Hominids of intermediate brain size were present, as is evident from the recent discovery of such a skull, of the same age, with a capacity of more than 800 cc, somewhat greater than that of the Olduvai specimens, which as we shall see are over half a million years later in time. We still know too little, but we should not hold to the preconception that tools appeared after a brain "large enough" had evolved. Instead, we should accept the fact that the making of simple tools gave the brain, in coordination with the hands, a selective milieu for reorganization and for cortical expansion.

In this and other connections I might remark that perhaps it is easy to lock onto a too simple hypothesis of brain-hand-tool relations in evolution, and by extension to adopt the chimpanzee as the model for an approximate ancestor in every way. Unquestionably he (with the gorilla) is our closest relative, from much evidence. He is largely terrestrial, stands erect with some ease, and has been observed making "tools" in the wild. None of this is true of the orang utan, whose hand is also distinctly less suited to manipulation, being specialized for brachiation, with a short thumb and

long, somewhat hooked fingers. By our simplest
hypothesis, above, he should be a less intelligent
animal than the chimp. But his brain is equally
large. And, in spite of hand form, his manipulative ability and curiosity with objects are high
(Jantschke, 1972; I thank Peter Rodman for this
reference). Although more phlegmatic in attacking
laboratory problems, orangs in direct comparisons were found to be more penetrating and quicker
in their solution than chimpanzees or gorillas:
three orangs solved one problem in two, three,
and 20 minutes respectively; one chimp solved it
also in two minutes but all the other animals took
several days or failed completely.

Furthermore, R. V. S. Wright (1972) has in-
duced a young orang to make flint flake tools by
striking them from a core with a hammerstone
This was done strictly via the monkey-see-monkey-
do route. The inducement was having a tool to cut
a cord to open a box with fruit in it, a step learned
beforehand. This is not very surprising as to mental process or brain size, given what is already
clear about capacities of the large apes; more
striking is the orang's ability to manage the work
at all, given his thumb and fingers. (He held the
flake with the thumb against the side of his index
finger, all fingers being folded into the palm.) On
the other hand, he used his prehensile feet to
steady the core while he struck it; we may reflect
that early hominids lacked this particular advantage.

From all this, we may conclude that, for pongids, tree life may provide at least as much of an
"enriched environment" as that on the ground, and
we should remember to ask ourselves how the

large apes attained their own level of intelligence
and brain size as part of the problem of final human progress. That is to say: the orang and the
African apes have been separate phylogenetically
for longer than we and the African apes; orang and
chimp have developed superior primate intelligence
in parallel; we must assume somewhat different
evolutionary situations for the two, and therefore
for ourselves. So we should not automatically assume single or narrowly defined causal factors for
brain progress either up to the australopith stage
or beyond it.

Returning to brain volume, we must be cau-
tious in considering this as the only sign of progress, since a reorganization emphasizing other
brain components might be masked by a simple
measure in cubic centimeters. As Holloway (1972)
says: "comparisons based on cranial capacities
alone are not comparisons of equal units...."
Another element of uncertainty is lack of real
knowledge of body size, since an australopith with
a brain the same size as a chimp's but a body only
half the weight would be relatively large-brained.
For the South African forms, estimates as to the
difference between Australopithecus and Paranthropus have varied, some making the former
really small (40 to 50 lb) and the latter perhaps
larger than modern man. Other comparisons see
the difference as much less. The evidence is not
really clear, since there are so many uncertainties
as to differences of sex and of individual variation,
as well as doubt in assigning parts of the skeleton
to one species or the other (see pages 28 and 32).

In addition to difference in size of brain,
which favors Paranthropus in absolute figures for

the South African australopiths, Holloway finds
evidence that the two species differed in shape and
proportions affecting different parts of the brain.
In any case I have not yet fairly introduced the
"gracile" form at Olduvai Gorge, contemporary
there with Paranthropus: the finds grouped by
many authors under "Homo habilis." In this population, which was apparently of quite moderate
body size, at least in females, the cranial capacity
of three individuals carefully evaluated by the most
qualified workers (Tobias, Holloway) suggests an
average between 650 and 700 cc. The still earlier
new skull (ER 1470), found by Richard Leakey in
1972, is estimated by him at over 800 cc. This
clear increase, relative both to the South African
Australopithecus and to Paranthropus, brings us
to the question of evolution within the australopithecine stage. (In fact, to consider ER 1470 as possibly within this stage, as I am doing here, is an
injury to the ideas of Louis and Richard Leakey,
as we shall see below.)

EVOLUTION IN THE
AUSTRALOPITHECINES: THE PATH TO HOMO

We have the beginnings of a chronology for evolution in the australopithecines, still shaky but, as
the result of much recent East African work by
paleontologists and geologists as well as anthropologists, much better than in the past. Pliocene
australopiths are known from the Lothagam mandible dated at over five million years ago, and the
Kanapoi humeral fragment at over four million;
but these have little else to tell us. From at least
three million, however, there is evidence in the

northern Omo-Rudolf region of the presence of two
distinct lines. Over the time range of 2.6 to about
2.0 million years in East Rudolf there are good
specimens of male and female Paranthropus skulls
plus the quite different ER 1470, as well as lower
jaws approaching those of Homo in shape. The important Olduvai sequence of Bed I and lower Bed II
is now thought to fall within the span of 1.9 to 1.6
million years, perhaps well within it; and the sites
in the north again carry hominid remains of more
than one kind down to about 1.3 million years or
less.

In South Africa, where australopith finds were
made so much earlier than in East Africa, the
sites have never been dated absolutely. It has
been agreed, on the evidence of fauna, that
Makapan (Limeworks) and Sterkfontein (both Australopithecus sites) preceded Swartkrans and Kromdraai (Paranthropus sites). The agreement also
rested partly on supposed correlations with general cycles of rainfall, now doubted, which indicated that the sites straddled the beginning of the
Middle Pleistocene, a relatively late date for
these sites in relation to the Olduvai sequence;
some estimates made them only about half a million years old. Recent opinion (Cooke and Maglio,
1972), however, based partly on evolution within a
baboon genus (Simopithecus, allied to the gelada—
Maier, 1972), now views the South African cave
deposits as very much earlier than previously,
with Swartkrans and Kromdraai not later than Bed
I at Olduvai, and the Australopithecus sites going
back to something like 2.5 to 3.0 million years,
and these figures may be conservative. With such
a revision the timing of stone tools in East and

South Africa, as well as their relative progress in development (M. D. Leakey, 1970a, 1970b), would fall into place much better.

The hominids themselves would also be put into a better arrangement, conforming generally to one long advocated by Robinson. Paranthropus, we may decide, remained essentially stagnant in his specialized, though completely hominid, adaptation from his first appearance (three million years ago, probably considerably more) down close to one million at sites in East Africa, where one of his recently found fossil jaws may be the most massive yet known. Australopithecus of South Africa, though fairly large-jawed to begin with (at dates around three million years), was soon exhibiting change and progress. Tools, present by 2.6 million years in the East Rudolf area and probably at Sterkfontein, can reasonably be credited to this lineage (and not to Paranthropus). By that time, on the evidence of ER 1470, he had certainly reached the stage of "Homo habilis"; and the somewhat later tools at Olduvai appear to show signs of technical progress.

Homo habilis, [11] of Bed I and lower Bed II at Olduvai (1.9 to 1.6 million years ?) was originally named from teeth, jaws, and cranial fragments, principally of the individuals OH 7, the "pre-Zinj child," and OH 16, "Olduvai George." The difference from Paranthropus (here in the guise of

---

[11] I have in most places put *"Homo habilis"* in quotation marks because of my own feeling that a full species may not be necessary for the transitional phase between *Australopithecus* and *Homo* (at least in the phylogenetic schema used here), and also because its original sense (Leakey, Tobias, and Napier, 1964) was in definition of a species which was not thus transitional at all, but outside of any direct line from *Australopithecus* to *Homo erectus*.

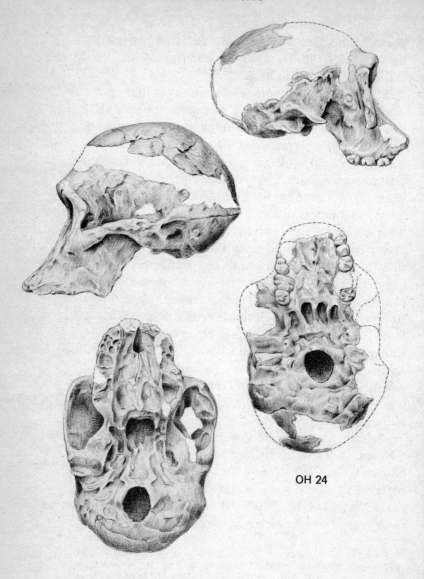

OH 24

Sterkfontein 5

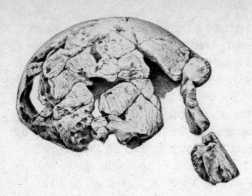

ER 1470

SKULLS OF *AUSTRALOPITHECUS—HOMO HABILIS*

This figure compares Sterkfontein 5, from South Africa, with OH 24 ("Twiggy") from Bed I at Olduvai, which is probably one to two million years later. There is a general similarity on conformation, but in OH 24 the brain is larger, the face is shorter, the teeth are evidently smaller, and the foramen magnum at the base of the skull is placed further forward.

ER 1470 is also shown. This is clearly larger-brained than OH 24, but also larger in the face and upper jaw. The fragmentary nature of the original find has allowed a good reconstruction of the braincase but has so far prevented a definite positioning of the face relative to the rest of the skull.

"Zinjanthropus") was clear. The teeth had some
kinship with known Australopithecus as to proportions: compared to Paranthropus the molars,
though long, were not so massive and the jaw was
relatively more generous in front, as well as less
heavy. Skull parts indicated a thin, high, round
skull. The most recent restoration of a skull from
the bottom of Bed I, known as OH 24 ("Twiggy"),
is perhaps the best Olduvai representative, and
suggests a general homogeneity among the "habilis"
remains. Although restoration was very difficult,
and leaves some questions of correctness, this
skull bears certain comparisons with the best of
the South African Australopithecus, i.e., Sterkfontein 5, or "Mrs. Ples." Especially if allowance
is made for possibly too great a flattening of the
vault in the restoration of OH 24, there is a general
likeness in form, though with the latter having a
brain case relatively and absolutely larger.
Viewed from below, it is seen that the individual
molar teeth, and the tooth row, must have been
shorter in OH 24 than in Sts 5, so that the face,
though having a similar concave profile, is less
projecting at the mouth. And the foramen magnum,
marking the point resting on the spine, is further
forward under the skull. Though this is a comparison of only two specimens, the differences are
highly suggestive of what evolution, in tool-makers
of the Australopithecus lineage, could have effected
during this period of development.

     The ER 1470 skull, certainly earlier and cer-
tainly larger-brained than OH 24, would then have
to be fitted into this line. While on a first look it
seems to have the appropriate facial characters,
it is large: its brain, estimated as not less than

800 cc, overlaps in size with Homo erectus. If it
is to be considered a "habilis," in the Australopithecus developmental line, then it would have to
be a male, with males and females being decidedly
dimorphic in size, although Olduvai habilis specimens approach 700 cc in brain size. It should be
emphasized that its differences from other habilis
specimens are indeed size, not shape as with
Paranthropus; the upper jaw indicates that the
front teeth (which are broken off) were quite large
and also broadly spaced, exactly the opposite from
Paranthropus.[12]   A supposition of sexual dimorphism (though in other applications a matter of
controversy—see below) is not pure shadow, because skeletal parts from Sterkfontein in South
Africa show surprising differences, some suggesting pygmy body size and some not (see above); and
the four-million-year-old Kanapoi fragment of a
humerus is just at the modern average for size. It
might be mentioned that, while the Makapan Limeworks site of Australopithecus has yielded no complete skulls, cranial parts suggest a braincase
rounded as in habilis or ER 1470, though with a
much smaller brain.

    Another item of evidence bears on all this.
Swartkrans has tools. They are not of the earliest
type, and the date is also not the earliest, being
perhaps equal to Bed I at Olduvai, i.e., around
two million years or a little more. Swartkrans is
a Paranthropus site—are we wrong in denying him
tool-making? Long ago, among the many

---

[12] I thank Richard Leakey who, with the generosity characteristic
of his family, has allowed colleagues to inspect and comment on
material before detailed publication.

Swartkrans fossils, Broom and Robinson found a
few fragments, including a lower jaw, and a mid
upper jaw piece, which they recognized as a different, more advanced hominid; they named it "Telanthropus." Robinson later simply put it down as
Homo erectus, accepting the presence of this more
advanced form as the explanation of the tools.
Twenty years after, while R. J. Clarke and Clark
Howell were looking over the Swartkrans collection
in Pretoria, the former noticed that parts of a
skull (SK 847), previously assumed to be Paranthropus, also appeared to differ from that form (Clarke,
Howell, and Brain, 1970). The pieces were put
together, and another connecting piece, a temporal
bone, was found and added. At this point it was
discovered that there was a perfect join to the original upper jaw fragment of "Telanthropus," showing that everything belonged to one individual. In
addition, it was clear that none of the deep jaws
belonging to the high-faced Paranthropus would
properly fit the new half skull, but that the "Telanthropus" lower jaw, with its low ascending branch,
was a mandible of the right form (if not of the same
individual). Other features, such as a more overhanging brow and a less flat face, are quite unlike
Paranthropus but suggest something more advanced
toward Homo erectus. The authors stop short,
rightly I think, of calling the specimen erectus,
but its suggestion of an intermediate form, perhaps
already beyond OH 24, is compelling.[13] At any
rate, we have evidence, here as in East Africa, of

---

[13] It might be noted that, compared to either *Australopithecus* (of
South Africa) or *Paranthropus,* both OH 24 and "Telanthropus"
have a more vertically pitched alveolar region of the palate (behind
the front teeth), which might be important in human speech; they

a progressive line contemporaneous with the
bogged-down Paranthropus.

Various controversies surround much of this,
though some are not essential. One school holds
that only a single species of australopith existed,
the idea of two forms being an illusion created by
sexual dimorphism, largely through the accident
of comparing large, rugged males from one site
with gracile females from another. This view
would have to overthrow a great deal of evidence,
some detailed above but also including such things
as the finding of a convincing female cranium of
Paranthropus (ER 732) which is not like Australopithecus, and which is contemporary not only with
a fine male Paranthropus skull (ER 406) but also
with some distinctly more Homo-like jaws, in addition to being later than the Homo-like ER 1470.
At the opposite extreme Louis Leakey and some
associates argued (L. S. B. Leakey, Tobias, and
Napier, 1964) that Homo habilis is not an australopithecine derivative but is distinct from both australopith lines (Tobias, 1966), forming a separate,
third group, which led eventually directly to Homo
sapiens, bypassing the later, thick-skulled Homo
erectus altogether. Leakey also suspected the
presence of still another, even more progressive
species, a real incipient Homo sapiens, something which the new ER 1470 would sustain better
than other evidence yet found.

It is safer not to multiply hominid lineages
more than the fossils strongly insist. In the middle of all the above is the view adopted here:

---

also are more *Homo*-like in the lower border of the nasal aperture
and the more forward position of the nasal spine relative to the border. (See Robinson, 1953b.)

Australopithecus and Paranthropus stem from an
earlier speciation in the hominids, with the former
leading to Homo and the latter becoming extinct.
Robinson, who has long held this view, would in
fact simply drop the genus Australopithecus entirely, putting fossils of the lineage root and
branch into Homo. Leakey, Tobias, and Napier
(1964), in their different interpretation, expanded
previous definitions of Homo so as to allow the ad-
mission of Homo habilis. Both suggestions are
probably relaxing things too much, though it does
not greatly matter. There was, of course, a continuous evolution, and the East African "graciles"
still seem rather small-brained (except ER 1470?)
and dentally primitive for Homo as described below. If a transitional species is accepted, it might
best be added to Australopithecus.[14]

In sum, we might use the following reconstruction for the present. The hominid ancestry divided
by speciation before five million years ago (I
frankly think the early Lothagam and Kanapoi fragments both represent Australopithecus). Paranthropus continued on the adaptive vector of a hard
vegetable diet to an extreme skull and tooth form,
but remained an imperfect biped. He represents
what, but for the other line, the hominids might be
today: cultureless bipeds fractionally brighter
than gorillas. Australopithecus advanced on all
fronts, as a better biped and as a more general

---

[14] If kept in *Homo*, *H. habilis* remains the name. If the Swartkrans
advanced hominid specimen (Telanthropus + SK 847) is included, as
seems proper at the moment, and if the taxon, or named group, is
put in *Australopithecus*, then the species would be *A. capensis*, by
reason of priority of the species name originally given to *Telanthropus capensis*, for the material incorporated in the specimen.

feeder whose fore teeth did not specialize down to
such small size. The South African cave fills,
perhaps around three million years in age, show
him still small-brained but well advanced in bipedalism. Before 2.6 million years either his
descendants, or more rapidly evolving East African populations of the same complex, had advanced
considerably in brain size; and interest in eating
dead animals had brought them to the point of making sharp little stone tools to cut through hide.
The find of a relatively large-brained skull of this
date, as well as two femora of Homo-like proportions, has been a surprise that the Olduvai habilis
specimens, later than two million years, had not
quite prepared us for. But the surprise is not total
and merely forces us, I believe, to expand our
ideas as to this proto-Homo phase. With its evidently large upper jaw, I conclude for the present
that ER 1470 should be accepted as a large, and
large-brained, male of this Australopithecus-Homo
transition. Other skulls of what I assume to be the
same population ("habilis") run to nearly 700 cc
brain size, and the differences presented by the
new skull (something over 100 cc) would not be
startling to find within the variation of gorillas and
can easily be matched in that of such early men as
those of Java (see below).

Also, lower jaws of an almost Homo aspect—
some of them looking indeed even more advanced
than known Homo erectus, though this may simply
reflect generally smaller skull size—occur at various time levels in East Rudolf, alongside jaws of
Paranthropus. In fact, there seems to have been
a span of over a million years, following 2.6 million or more, which has so far yielded nothing

showing distinct progress beyond what ER 1470
represents. As far as Africa goes, from this point
(about 1.6 million years ago) we remain much in
the dark as to evolution until the positive appearance of Homo erectus in Africa, and of tools of the
Acheulean, hand-axe form. For one thing, dating
is still poor in this time interval. Nonetheless
there are a few signs, such as the OH 13 skull,
"Cindy," from higher up in Bed II at Olduvai than
the other "habilis" finds, and also some of the
East Rudolf jaws of apparently late date. These
things point to a Homo-like population, more advanced than "habilis" but smaller in body, skull,
and teeth than Homo erectus.

Homo Erectus

THE PROBLEM OF DATES AND ORIGINS

When Homo erectus (i.e., the conventionally recognized specimens) does appear, changes have
taken place. Cranial capacity is larger, though
covering a wide range (estimates are from 750 to
1225 cc for various crania sufficiently complete
for estimation). More striking are three other
things: (1) body size and form are not significantly
different from man's today; (2) the skull is much
thicker, broader-based, and relatively lower than
in Homo habilis; and (3) in spite of this the braincase—putting it impressionistically—now dominates
the face and jaws rather than vice versa.

There is thus a hiatus between this form and
the latest known australopithecines (habilis). It is
sufficient so that, for the present, we can look on
it as a new hominid stage. It was sufficient to
cause Leakey, Tobias, and Napier (1964) to deny
a lineal connection between habilis and erectus.
Is this correct? We are faced with a clear degree

of difference but also with an indefinite interval during which that difference could have come about by direct evolution. For the first, the absolute difference in body size suggests that advance in brain volume was actually relatively trivial or even nonexistent. In this case the cruder, lower skull of erectus, looking like a fallen soufflé in comparison with that of habilis, simple represents growth in which skeleton outstripped brain, so that the bony envelope containing the latter indeed has a fallen look. For the second, the time gap in the fossil record, we recognize that there is still much uncertainty as to the earliest dates for Homo erectus.

No specimens are known from before the Middle Pleistocene (perhaps corresponding to the faunal change in Bed II at Olduvai), although the beginning of this is not very well dated in actual years. The Olduvan erectus braincase from upper Bed II can only be said to be older than 700,000 years, on the evidence of paleomagnetism for the base of Bed IV (M. D. Leakey, 1971). Java furnishes the crucial material for first developments, erectus specimens having come from two faunal zones, the Djetis and Trinil, in the following sequence:

Geological divisions	Faunas	Pleistocene assignments
Notopuro	Ngandong	Upper Pleistocene
Kabuh	Trinil	Middle Pleistocene [See
Putjangan	Djetis	Note 1 on page 160 (Sino-Malayan fauna
Kalibeng	Kali Glagah	Lower Pleistocene
Kalibeng	Tjidjulang	(Siva-Malayan fauna = Villafranchian)

Datings for the Trinil faunal zone cover a wide range (Isaac, 1972). Two dates on pieces of basalt, believed to be from its higher levels, are just about 500,000 years. Tektites[15] from the Australasian fall, an event well dated on the tektites themselves at about 700,000 years, have been found in the lower part, a date supported by more recent ones on volcanic tuff (von Koenigswald and Ghosh, 1973) averaging 830,000 years. Still another, on pumice from the locality where the second Java skull was found in 1937, gives an age (with very wide margins of error) of 900,000 years. As to the Djetis zone, there are two dates, both of about 1.9 million years. One is from a level well below human fossils but the other is from Modjokerto, stated positively to have been fairly close in level and locality to an erectus child's skull (Jacob and Curtis, 1971).

Apart from this child, there have come from the Djetis zone parts of an adult skull (IV) and a separate lower jaw (B). The rest derive from the Trinil, except for Pithecanthropus III, which may have come from the base of the Notopuro series just above (von Koenigswald and Ghosh, 1973). Thus a really considerable time span has now been sampled. The very latest skull may date from less than 500,000 years; the date of the earliest should be over one million—on face value, almost two million. Such a great age as this, however, must for the present remain suspect. Not only would it make the span of the Java erectus people almost a million and a half years. It would also

---

[15] Tektites are glassy nodules obviously heated by falling through the atmosphere in swarms. There is still argument as to whether they derive from space or from terrestrial volcanic explosions.

set early Java man as contemporary with the earliest "habilis" individuals at Olduvai. Hence one interpretation would be that hominid evolution in Asia was far in advance of that in Africa, so that a westward migration of Homo erectus would allow for an abrupt appearance of Homo erectus in Bed II at Olduvai. The other interpretation is that early dates for man in Java need confirmation, and this is the obvious choice. A time span for Java man of half a million years or something over would be more appropriate for the visible changes there in both hominids and animals. And a maximum age closer to one million than two million years would be more in keeping with what is estimated for the start of the Middle Pleistocene elsewhere.[16] It would also leave open a moderate time gap, following about 1.6 million years, for the final evolution of "habilis" into erectus.

As things have been developing, there appears to be no possible source for Homo other than the basic Australopithecus line. Further evidence from Java bears obliquely on this. Along with recognizable specimens of Homo erectus, there have been recovered from the Djetis zone four or five jaw fragments, with teeth, of another hominid, originally named Meganthropus by von Koenigs-

---

[16] In Europe the beginning of the Middle Pleistocene is currently dated somewhat later, with the Cromerian fauna of the Günz-Mindel (or first) interglacial, at about 700,000 years (Cooke, 1972). Such discrepancies remain to be explained. No signs whatever of early man had been found in Europe before the Middle Pleistocene until the de Lumleys discovered small chopping tools in Villafranchian deposits in a shelter, Vallonet, on the French Riviera.

wald.[17] Bulkiness of the jaw and massiveness of
the molar teeth betray an australopithecine.
Robinson (1953a) holds that it is in fact recognizable as Paranthropus; the one specimen with the
forepart of the mandible, though crushed, gives
signs of the small canines in combination with robust cheek teeth which are characteristic of that
form. Von Koenigswald (1967) agrees that it is a
probable australopithecine, and reports he has
found (in Chinese apothecaries) teeth from China
which are also probably australopithecine. If this
is the correct interpretation—and little better than
"if" is possible now—then we seem to have one
more example of Paranthropus surviving as a contemporary of the line evolving from Australopithecus into Homo, as at Olduvai, at Swartkrans, and
in the East Rudolf region.

A special point: we cannot make decisions as
to which continent was the "cradle of man," important as this seems to be in the public mind.
From the many biological ties, especially biochemical, between man and the African apes, it is in-
escapable that Africa was the scene of the separation of hominid from pongid ancestry. However,
before 10 million years ago Ramapithecus was already distributed from Africa at least to India (and
probably Europe and China—see Simons and Pilbeam, 1965); and we see Homo erectus in Africa
and Southeast Asia apparently equally early. The
evidence is slight, but australopithecines were
probably likewise present in both; different ad-
vances might have occurred in different places.

---

17 I thank Professor von Koenigswald for some of the information
and for showing me original specimens.

EVOLUTION WITHIN HOMO ERECTUS

In the Far East the relatively numerous specimens
of the Java and Peking men allow us some view of
change over the time span we have provisionally
adopted, the first half million years of the Middle
Pleistocene. Javanese discoveries have been
sporadic. The first was Dubois' sensational find
of 1891—Pithecanthropus, the ape-man—in silts of
the Solo River at Trinil. The others have come almost entirely from villages around the Sangiran
Dome in Central Java, a volcanic uplift which has
been eroded so as to expose, at the center, marine
clays of the Pliocene and, further away, in con-
centric exposures, beds containing the Djetis fauna,
then the Trinil, and then, around the whole, deposits of the Upper Pleistocene. Thirty-five years
of searching, first by von Koenigswald and later by
Indonesian scholars who have been his students or
colleagues, have made villagers all around Sangiran into amateur anthropologists, and discovery
promises to get better all the time. A total of
eight or nine skulls is the bag so far, with various
additional fragments.

Only one adult skull, lacking the frontal region, is associated with the Djetis. It is particularly low and thick, and has an estimated internal
volume of 750 cc. Like all the Java specimens it
has a well-defined, thickened mound or torus of
bone across the back marking the attachment of
the neck muscles, this torus also forming an acute
angle between the upper slope behind the low crown
and the lower part of the back, the nuchal area.
There is a particular lumpy thickening or eleva-
tion along the midline—not a sagittal crest like

Paranthropus at all, but something which Weidenreich considered a sort of structural reinforce-
ment, though it seems thick beyond imaginable
functional needs. The specimen has a large and
long palate. A separate lower jaw seems to cor-
respond well to the upper jaw in size and characters. Both lower and upper molars are longer
than modern man's but well reduced compared to
the early australopiths. The incisors are broad
and well spaced; the fore teeth in fact seem relatively more important than in australopithecines,
especially the "robusts." There is even a slight
gap between the upper incisors and canines, seen
in extra fragments as well, a feature contrasting
greatly with the narrowing and crowding of front
teeth seen in Paranthropus.

The Trinil specimens mostly lack teeth or
facial parts. Though we are comparing them with
a single earlier skull, they seem to exhibit reduced robustness. Males and females appear to be
represented; one or two of the former have the
same lumpy thickening along the midline. In addition they have a slight depression running fore and
aft along the skull on either side, as if the roof
were a little deflated between the raised midline
and the angles at each side where the sides of the
skull turn down to slope more vertically to the ear
region. Presumed females lack this midline ele-
vation but have the same outline in softer form.
They have the thick but smooth torus making an
angle at the back; one male, however, has a downward hanging crest instead.

The average brain volume for the best estimates of five Trinil skulls (Tobias, 1971a) is 860
cc. Considering that this must include females, it

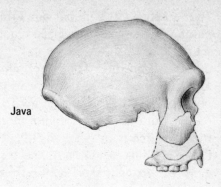

Java

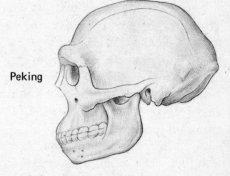

Peking

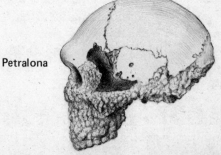

Petralona

SKULLS OF *HOMO ERECTUS*

The Java skull is drawn from a photograph of "Pithecanthropus VIII" by Sartono. The upper jaw, displaced in the original specimen, has been repositioned here to a probably better relation.

The Peking skull follows Weidenreich's restoration. The Petralona skull is from a published photograph; adhering material has not been removed from much of the face.

seems like a clear increase over the lone Djetis male with his estimated 750 cc (see Note 2 on page 161). Furthermore it does not include the provisional figure of 1029 cc for the most recently reported find, number VIII (Sartono, 1971). This fine skull, still under preparation and study by Dr. Sartono, has much of the face in undistorted position. It is broad and robust, but carries no suggestion of the deep faces of Paranthropus. The palate and teeth appear to be distinctly smaller than those of the Djetis specimen. A few other facial fragments, especially cheekbones derived from both Djetis and Trinil, suggest faces less robust than this one, and in fact seem surprisingly modest in size. But little can now be said about the whole conformation and projection of the face.

A single skull, and a separate jaw, from Lantian, South China, are believed to be of early Middle Pleistocene date. In form the finds seem to be on a level with the Djetis Java man. Peking man of North China provides us with a whole fossil population. This was represented by numerous skull and jaw fragments, and a few skeletal parts, which were dropped to become fossilized in the

large cavern, Locality 1, in limestone bluffs at
Choukoutien. Although the remains came from
many levels in the fill, this does not seem to mean
important time differences. As in Africa, dating
the contents of such a cave has been a problem but
the present estimate is a warmer interruption in
the second major glacial phase (the Mindel of the
Alps). A preliminary estimate of absolute age,
made directly on animal bone from Locality 1 by
the amino acid reacemization method, is 300,000
years.[18]

Such an age is appropriate, if somewhat less
than previous estimates. It would make Peking
man anything up to a quarter of a million years
later than the latest Javanese fossils. Progressive evolution over this interval might allow the
men of Peking to be direct descendants of the Java
form, as far as most skull features go. The main
general difference is the greater size of the brain:
from five specimens Weidenreich estimated the
capacity of the braincase to run from 915 to 1225
cc, averaging 1043. Capacities of the Trinil
skulls, we saw, appear to have fallen in a range
from 775 to 975 (Tobias, 1971a), with an average
of 860. The greater volume of the Peking brains
is reflected in a higher skull, showing a forehead
angle in profile view, and with a lack of the "deflated" look on either side of the midline seen in
at least two (male ?) Javanese crania. The side
walls tilt in less abruptly over the ears.

At the back, the whole breadth of the occiput
is less. But the torus, or thick mounding at the

---

[18] The basic method is given in Bada and Protsch (1973). For this
unpublished preliminary date I thank Drs. Bada and Protsch, and
Professor von Koenigswald, who supplied the bone sample.

top of the neck muscles, is equally prominent,
though it is not known to have formed a sharp overhanging crest in any case. The angle in profile is
equally acute. And, in general, the basic form is
similar, appearing only to be better filled out.
Weidenreich found the bone of the skull to be at
least as thick in the Peking as in the Java men
(both forms being much thicker than modern
skulls). Some of the newer finds of Java man may
turn out to be thicker than specimens found earlier,
which are now thought to be mostly those of women;
this may redress average thickness in favor of
Java man. Still, in the Peking skulls at least,
thickness is real heaviness, a thickness of the
dense bone of the outer and inner tables (surfaces),
not a general expansion, especially of the spongy
bone between these tables. The same greater-
than-modern thickness is found in bones of the
face and of the body.

In addition, the bony brows continued to be
heavy, and in fact contained only small frontal
sinuses of air hollows in the region just over the
nose, being mostly solid bone. Sinuses in the Java
skulls may have been larger, but probably not
much. The face below was complete in no Chinese
specimen, but enough fragments were found to give
reliable information as to the whole. There was
no indentation at the top of the nose (a slight Gre-
cian effect) but the nasal bridge was flattish. The
midface was neither recessed nor prominent,
though pulled back a little under the prominent
brows, while the mouth region, on account of teeth
larger than modern man's, was protruding. The
chin region was correspondingly withdrawn, though
still forming a good angle, rather than a smooth

curve, with the lower border of the mandible. On
the whole, the face was broad and short, flaring at
the corners of the jaws. Comparison with Java
man is not useful at the moment, beyond recalling
that the latter's face, while broad, was probably
neither long nor relatively particularly large, compared to the australopiths.

The fairly numerous Peking teeth and jaws allow us to confirm the remark that the molar teeth
in Homo erectus were much diminished from the
australopithecines. In the following table I have
taken figures (partly from Robinson) for the average length and breadth of lower teeth, and multi-
plied the two together, as if each tooth were rectangular, to give a crude comparison of surface
size. (The figures are in square millimeters.)

	Incisor 1	Incisor 2	Canine	Pre-molar 1	Pre-molar 2	Molar 1	Molar 2	Molar 3
Swartkrans (Paranthropus)	34	45	62	113	147	210	243	244
Sterkfontein (Australopithecus)	43	54	99	107	117	178	313	208
Peking men	41	48	81	86	92	149	150	131
Modern U.S. white	32	38	55	53	57	115	108	105

This will show that the Peking teeth, while
distinctly larger than modern man's, are larger
more or less in proportion; also, the third molar,
or wisdom tooth, is distinctly diminished compared
to the second, as in Australopithecus as well.
Peking man's anterior teeth, however, stand comparison with Australopithecus from Sterkfontein,

and entirely overshadow those of the "robust"
Paranthropus from Swartkrans. It is in the five
cheek teeth that all the differences occur. As we
have seen, the size in Paranthropus bounds up behind the canine tooth, from 62 square millimeters
to a first premolar with 113, and then to a second
premolar of 147, while the jump in Australopithecus, as in Homo, comes at the first molar. The
Peking figures show that the whole cheek row has
dropped off in size, much more than the incisors
and canines.

GEOGRAPHICAL DIFFERENCES IN HOMO ERECTUS

Remains of skulls or jaws assigned to Homo erectus in the west are scanty. Olduvai Gorge has
yielded one skull top, from the upper part of Bed
II (i.e., with a new fauna and with more advanced
tools, Acheulean hand axes), and dated before
700,000 years, perhaps well before. Louis Leakey
found it in 1960 but, in the press of many other
important discoveries, it has not had its chance to
be described in detail by an anatomist. Its brain
volume has been estimated at 1067 cc (Holloway,
1973), and its probable date would make it roughly
contemporary with Java man. In profile, however,
it has little forehead and it is embellished with
brow ridges more voluminous and flaring than
either the Peking skulls or the Java skulls; and
these brows have a slightly different look. The
skull is said to be thick.

From an animal fossil site of the same or
earlier date, Ternifine, near Oran, Algeria, Pro-

fessor Camille Arambourg extracted (from 1955 on) three human lower jaws, and the parietal bone (one of the pair forming the middle of the skull roof) of a young person. He named his kind of man "Atlanthropus," but not in a serious attempt to separate him from the Homo erectus group generally; certainly there is nothing in the Ternifine pieces to justify such separation. The molar teeth are somewhat larger than those of Peking man. The parietal bone is less thick than those from Peking (a question of youth?), though thicker than modern man's; and the shape of the bone's edges indicates the same kind of low, nonglobular braincase already known from Java and Peking. Internal markings for blood vessels are also of a primitive pattern.

Europe has also been stingy with human remains of the time, providing two, or at most three, fossils. One is the occipital bone (back of the head) found in 1965 at an early man's camp site west of Budapest, at Vértesszöllös. The piece was alone (i.e., detached from the rest of the skull before fossilizing), though two tooth fragments were also discovered at the place, which can be placed rather definitely in the milder interval during the second (Mindel) glacial, the probable time of the Peking cave. This piece is primitive, with a wide, flat region for neck muscle attachment which forms a marked angle, like other erectus skulls, with the upper half of the bone sloping back from the crown; however, this upper half is well curved as in modern man (Thoma, 1972). The bone is also fairly thick. But the torus defining the angle is not comparable in thickness or consistency with the Peking or Java skulls. And A. Thoma (1966, 1969), in describing the fossil, did computations (regression equations) on the relations of brain size to measurements and shapes of the occipital bone in different kinds of complete human skulls, leading him to a calculated estimate for this individual of 1350 to 1400 cc. Now this is entirely in the range of modern man—in fact, between male and female averages—and Thoma decided that the Vértesszöllős individual would have to be recognized as falling on the border between erectus and sapiens (insofar as there is a border).[19]

The lower jaw from Heidelberg (or Mauer, the village of its actual discovery) was found long ago, in 1907, and at that time impressed scholars with its robustness. This is still impressive, though not when viewed beside the mandibles of australopiths, or of Meganthropus. Nor are the molar teeth particularly large, though their deep crowns exceed modern ones in volume. The profile of the jaw differs from those of Peking and Ternifine (which are mutually rather similar), being more robust and deeper in the fore part, and exceptionally wide in the ascending branch; and having a rounded, receding chin instead of a more abrupt angle, as in the Peking jaws. What these differences of shape mean is not at all clear. The sands in which the jaw was found contained a fauna generally assigned to that which followed immediately on that of the Lower Pleistocene, the Cromerian of the earliest Middle Pleistocene of Europe, suggesting a date of 700,000 years (Cooke, 1972). Most therefore agree in placing the fossil in the

---

[19] Thoma's figure and conclusion have been questioned by Wolpoff (1971), who has been answered and rebutted by Thoma (1972).

first, or Günz-Mindel, interglacial, though both earlier and later assignments have been suggested. It is certainly an early fossil.

This miserable collection for Europe—one jaw, one occipital—has apparently been handsomely augmented by the discovery, in 1959, of the beautifully preserved Petralona skull in a cave in Greece. In first reports it was confused with Neanderthal skulls and assigned a late Pleistocene date, but recent work on the cave fauna, not completely reported, indicates that it is not later than the Riss Glacial, and probably much earlier (i.e., pre-Mindel again), which would put it well back in the Middle Pleistocene, perhaps coeval with the Heidelberg individual. The skull is large but rather small-brained (1220 cc).[20] Like recognized erectus skulls, it slopes in rather sharply from a broad base, and the profile of the rear is well angled. Hemmer (1972) has shown that relations of brain size and skull length are those of other erectus crania, conforming to its external features. It is still waiting to be cleaned off and properly described, but from its size and form it appears to rate as a moderately advanced specimen in the whole erectus spectrum, and one differing from the Olduvai and Far Eastern examples (though it is possible to imagine, from pictures, that an expansion and "advancement" of the Java Trinil specimen VIII might produce something like Petralona). Actual study of anatomical details would tell us a

---

[20] The method of estimation has not been stated as far as I know. Other estimates, of 1440 cc and 1384 cc respectively, have been published, but the method used for the first of these is the dubious one of applying estimation formulas, based on outside measurements of the skull, which were originally developed for modern man.

great deal more about this very important skull.

At the moment, therefore, the western material assigned to Homo erectus remains scattered and unsatisfactory compared with the eastern. Could the Olduvai man have had the Ternifine jaw? Was there a significant degree of intercontinental variation, an important question? The Vértesszöllös man does seem to have differed from his contemporary, the Peking man. But we can say little. Sticking to lower jaws, T. D. Stewart (1970) finds certain differences between those of the eastern (Peking, plus one from Java) and of the western (Ternifine, Heidelberg) erectus men. The latter have a much greater distance between the mental foramina (where the large nerve emerges on either side of the jaw), as well as a greater width across the teeth; on the whole, the tooth row seems shoved forward, relative to the body of the jaw itself, in the western specimens.

BONES OF THE SKELETON

So much for the cranial parts of Homo erectus. All too little is known of the rest of the skeleton. The "erectus" species name survives from Dubois' original name for Java man (see page 4), which he chose because of the fully modern aspect of the thigh bone which he found at the same level as the skull cap, but 15 meters away. Years afterward several further examples—shafts with the ends missing—of other thigh bones, equally modern as far as visible details go, were found in Holland, in boxes of bones from the same site, collected later in the 1890s but never picked over. Thus a general modernity in body size and form appeared assured

for Java man. Recently, however, Professor
Michael Day and Miss Theya Molleson in London
became suspicious of these leg bones and tested
them chemically with new methods, but could
neither confirm nor deny an age equal to the skull's.
Professor Day points out to me that these beds of
the Solo River, at Trinil, are apparently not primary deposits (like those at Sangiran), but have
been laid down more recently, in typical fashion on
the inside of a bend in the river. This has probably washed in both remains of the Trinil fauna
from elsewhere (the skull included) and other material which could be considerably, perhaps very
much, later. The possibility does not deny the
likelihood of generally modern size for Java man,
but it does remove the absolute assurance of a
completely modern skeleton suggested by the thigh
bones.

Few skeleton bones were found at Choukoutien
(the Peking man cave), and none (except for very
small ones) were complete. They did, however,
show that the modern body size was a fact, and
that there were no differences from modern man
bearing upon purely human form or gait. There
were, however, distinctions in detail. In particular, the walls of the long bones were especially
thick: the internal canal was much narrower, even
though the external form was entirely similar to
modern man's. Other differences were very
slight: a bowing of the thigh bone no greater than
is frequent in modern man but placed lower down;
and some differences in muscle markings below
the hip joint. Small though these are, the absence
of any such peculiarities in the Trinil thigh bones
led Weidenreich, in 1941, to conclude what I have

said: that it was not justified to attribute them un-
questioningly to Java man.

Finally, in Bed IV at Olduvai, Mary Leakey in
1970 found a hip bone, in fairly good condition, and
much of the shaft of the thigh bone of the same individual (M. D. Leakey, 1971; Day, 1971). This
is of course higher up, and therefore later, than
the Bed II skull, but how much later cannot be
said; it has a guess date of 500,000 years (the base
of Bed IV being 700,000), and at any rate the bones
should be considered those of Homo erectus. They
are very robust. The hip bone (innominate) has a
vertical buttressing over the joint (iliac pillar)
which resists muscular stress in firming the joint
when the body weight is being carried on that side
in walking. The feature is typical of man, but the
development here is extraordinary. The thigh bone
is particularly flat from front to back in its upper
part, a trait also well marked in the smaller Peking femora. As with the Chinese skeletal parts,
the Bed IV bones are instantly recognizable as
Homo, but exhibit traits which, in combination,
would be very hard to match in modern man. The
suggestion is that further evolutionary modification
did occur after the erectus stage, particularly if
the thigh bones from Java are held out of consideration.

A SUMMARY FOR HOMO ERECTUS

We seem to have a "stage"; let us remember that
it is cut off by something of a gap in fossils before
and, we shall see, after. There is variation in
the fossils. Part of this we must accept as evolutionary in origin, conservatively putting the Far

Eastern population in an evolving progression from
the earliest Javanese to the latest Chinese. Part
of it, we should expect, is the natural tendency to
divergence among populations isolated from one
another. Fossils of the western Old World are no
more than scraps, but we must be careful to note
that the Heidelberg jaw would not look at home in
the Peking material, and that the jaws from Ternifine have some differences from both. As to other
bones, perhaps only size differentiates the Olduvai
and Peking legs; Day finds agreement between
them in several other special features different
from modern man.

Can we deduce any significant differences in
adaptation or behavior between the australopithecines (late) and Homo erectus (early)? Signs are
hard to see. The first increase in relative brain
size seems unimportant, if not nonexistent. Earliest Middle Pleistocene stone tools do not seem
much better in technical production than, say,
those from Swartkrans. But the tools are larger,
and so is body size. Since the tools were probably
butchering implements, there is implied some difference, from scavenged meat, or the small "slow"
game whose remains as food were found at Olduvai,
to the much more numerous large mammals hunted
by the later men. This might be the selective
factor for the larger body size which is so evident;
and the molar tooth reduction might also be related
to a diet of more meat. Possibly this marks the
point of man's embarking on his long career in the
active and efficient hunting of game, with whatever
else resulted in his physical and cultural development. We shall come back to this.

The Rise of Modern Man

THE MIDDLE-UPPER PLEISTOCENE HIATUS

From the end of the second glacial phase to the
middle of the third interglacial—two and a half
major stages in the glacial sequence, running perhaps from 400 to 100 thousand or fewer years ago—
there is nothing comparable to the Java or Peking
finds. In fact, for a long time it has been assumed
that virtually all the known fossils of this time
came from Europe or North Africa and, except for
jaw fragments, consisted of two skulls. The lec-
turer in an introductory course could pay obeisance
to "Steinheim and Swanscombe" and move on to
Neanderthal man.

One effect has been the kind of apparent hiatus
I mentioned in the opening pages: the comfortable
division of fossil Homo into early erectus and late
sapiens, a break in morphology as well as in time.
Coon (1962), however, seeking physical criteria
for the division, decided that two apparently late
forms should be classed as erectus: the Rhodesian
and Solo men. He raised a good question.

For these two fossil men the problem and the gap remain unspanned. But in Europe things are not what they were 30 years ago when, except for the two skulls just named, the Heidelberg jaw reigned alone in the early Middle Pleistocene. There is now fossil evidence strung through every major geological phase[21] from the first (Günz-Mindel) interglacial on, the difficulty being the incompleteness of almost all specimens. Let us review this train, before turning to Asia and Southern Africa.

1.   Europe and its borders

We have seen that the "erectus" specimens, while few, might be looked on as slightly advanced in morphology for their period. If Petralona is actually very early (pre-Mindel) this is certainly true: for ten years the skull was thought not to be erectus at all. The Heidelberg teeth are not notably primitive. And Thoma named the second glacial Vértesszöllös bone Homo erectus seu sapiens in reluctance to assign it to either species outright.

The best skulls came from the late part of the second (Hoxnian, or Mindel-Riss) interglacial, about 250,000 years ago. One is from gravels of the Thames at Swanscombe and one is from

---

[21] Here I am using the old series of four "glacial" phases, purely for simplicity and familiarity of reference. These are the Günz, Mindel, Riss, and Würm glaciations of the Alpine sequence, discerned at the beginning of the century. There are reservations to be noted today: (a) the succession of warm and cold phases is now known to be more complex; (b) milder glaciations extend much further back, and the Günz of Europe falls in the latest part of the Villafranchian stage, or the Lower Pleistocene; (c) intercontinental correlation of climatic stages is unsure, to say the least, except possibly for the later Würm.

Steinheim, near Stuttgart, Germany. Both are
probably female. The Steinheim specimen is low,
and has puffy brow ridges, with a brain volume
somewhere between 1100 and 1200 cc, which is
smallish but, for the general skull size, respect-
able, and in any case above the known Homo erectus figures (Vértesszöllös excepted; Petralona is
a bigger skull). The face is rather straight,
tucked well under the brows, and is of moderate
size, as are the teeth. And at the back, the rounding of the profile and lack of anything like the torus
of the erectus specimens give a decidedly modern
look, in spite of the lowness. The back (occipital
bone) and sides (parietals) are all that remain of
the Swanscombe skull, and these parts look even
more modern, especially since the brain size was
at least 100 cc above Steinheim. It was long considered to be perhaps an early example of really
modern man, in spite of its marked resemblances
to the back part of the Steinheim skull. Studies by
Breitinger (1955), Stewart (1961), and Weiner and
Campbell (1964) nevertheless find primitive traits,
and indicate that the hind part does indeed resemble Steinheim, and that the missing fore part was
probably also like Steinheim, with enlarged brows
quite out of the modern range. Nonetheless these
two individuals are so decidedly different from
erectus that Coon (1962), on criteria of tooth and
brain size, and indices of curvature, rates them
as having crossed the admittedly arbitrary border
to Homo sapiens. At the same time Thoma (1969)
found Swanscombe and Vértesszöllös to be close
together in a multivariate distance measure, as
well as in many minor anatomical details, although
the general character of Vértesszöllös is more
primitive.

A good lower jaw from Montmaurin, near
Toulouse in southern France, is probably contemporaneous with the Swanscombe and Steinheim
skulls. It is not very big, but it is robust and suggests a projecting face; the molar teeth are also
generous in size. It has very faint suggestions of
the Heidelberg jaw in some features; in others it
is thought by some to foreshadow Neanderthal jaws.
At the same time, it has been suggested as a fit
for the Steinheim skull. However, obvious though
this suggestion is, the jaw creates a somewhat different impression of affiliation, especially in its
indicating a projecting face.

To the third (Riss) glacial belong skull parts
from the Abri Suard (of the La Chaise cave group),
Charente, France, which come from different individuals but are enough to make up most of the
back and side (occipital, parietal, and temporal
bones); there is also a lone parietal from the
Lazaret shelter, near Nice. Add to these an excellent fore part and face, and two lower jaws,
from the Arago cave in Tautavel near the Spanish
border of Mediterranean France, found by the de
Lumleys, and you have enough to make a "man"
of the French Rissian phase. (I say hastily that
this is no way to make a man, let alone a population from somewhat different parts of the Riss:
Arago is at the beginning, Suard is late.)

The pieces are surprising in seeming less
"sapiens" than their immediate predecessors, as
well as Vértesszöllös. The Suard cranial vault,
as far as its shape can be determined (Piveteau,
1970), appears low and flattish, though broad, and
the occipital is fairly sharply angled; all this is
reminiscent of Homo erectus. Thoma's method

for estimating cranial capacity from the occipital
bone gives a figure of 1063 cc, certainly low.
Other features of all the parts seem to support
Piveteau's suggestion that they represent forerunners of the Neanderthals, though preserving
some more primitive traits.

The Arago face would not disagree. The moderately flat forehead is not particularly narrow
behind the eyes, but the bony brows are strongly
developed, overhanging the upper nose and the eye
sockets, which are low and rectangular. In these
respects the face is not unlike Steinheim but does
not resemble that of the Neanderthals, nor has it
the great vertical length of the latter. However,
the lower nose and the upper jaw, as well as the
cheekbones, show a marked protrusion here. In
other words this face is more Neanderthal in
character below than above, and the Arago lower
jaws support this in part, but only in part. They
seem to lack the marked Neanderthal forward
placement of the teeth relative to the whole jaw
(see below); both specimens have characters like
the Heidelberg jaw (erectus!?), which is not as
large as the larger Arago mandible (H. and M-A.
de Lumley, 1971; I am also grateful to the de
Lumleys for other information and pictures). The
de Lumleys note resemblances to the Montmaurin
mandible as well, which also has slight resemblances to the Heidelberg jaw.

All these fossils from the third glacial make,
at least provisionally, an acceptable matrix of
transition from Homo erectus to the Neanderthals.
It might derive immediately from Steinheim, according to the Arago face, though Steinheim already seems more advanced and, like Swanscombe,

has been generally rated as sapiens in grade. The
comparatively gracile character of the Swanscombe
parts thus seems particularly surprising. Ultimate
ancestors could be the Heidelberg and Petralona
people; at least present knowledge would not refute
the suggestion. But none of this means that we
have recognizable early Neanderthals, like those
of western Europe in the Würm, only that there
were populations, greatly varied in character as
Piveteau (1967) notes, seeming to represent a
stage between erectus and Neanderthals proper.

In the ensuing third (Eemian, or Riss-Würm)
interglacial, which lasted from perhaps 150,000
to 70,000 years ago, the supply of fossils hardly
improves. Unsatisfactory parts of two skulls come
from the Fontéchevade cave, close to the La Chaise
cave group in the Charente. They are associated
with datable pre-Mousterian tools and a fauna establishing their age. (The Swanscombe woman, or
her husband, was a maker of hand axes of the long
Acheulean style, a basic tool also present at Ternifine and in Upper Bed II at Olduvai.) One of the
Fontéchevade specimens is a charred and somewhat distorted skull cap; it suggests a head moderately high though flattish in profile; the bone, however, is rather thick. The other piece, a small
bit of forehead, shows a definite absence of projecting brow ridges, and so gives an appearance
of modernity.

Whatever these really mean for earlier Homo
sapiens, a recognizably Neanderthal braincase
comes from Ehringsdorf in East Germany, and a
natural brain cast, also of evident Neanderthal
shape, from Gánovce in Slovakia. The only good
specimens are two skulls from Saccopastore just

outside Rome, and these show that the Neanderthal
skull and face had become established by this time.
We shall leave this for later discussion. However,
we must recognize, now as well as later, that if
Neanderthal man and modern man had any common
ancestors in Europe, these must be represented
among the scanty earlier fossils, and that in the
third interglacial the two lines of descent were apparently diverging, probably in different continents.

Foothills of the last glaciation. In the earlier part
of the Würm glaciation we meet a series of important finds suggesting populations which can be
distinguished physically from one another, but
which are all decidedly to be assigned to Homo
sapiens, and which are in fact rather modern in
basic characters. The two Jebel Ighoud skulls,
found in Morocco in 1961 and 1963, have rather
large brows and low skulls with projecting though
rounded profiles in the back. They have usually
been simply classed as Neanderthal, but I join
Professor Piveteau (1967) in rejecting this assess-
ment; at the same time, they are by no means
simply "modern." A particularly nonmodern trait
is the considerable breadth across the occipital
bone. But the face is generally modern in appearance, rather short, and recessed well under the
brows, though prognathic in the mouth region—evidently, the teeth were large. (The only known
permanent teeth--six-year molars in a child's jaw
from the site (Ennouchi, 1969)--exceed in size
those of both Neanderthals and modern men, having
a surface area of almost 180 mm.) Such people
could easily be direct descendants of Steinheim,
as far as all these features go. The culture here

is the Middle Paleolithic Levalloiso-Mousterian.
Radiocarbon dating (Ennouchi, 1968) from the site
gave results beyond the maximum age determinable
by the laboratory at Nancy, but this limit is not
stated. Projection from dates in Cyrenaica has
suggested about 45,000 B. C. (Oakley and Campbell, 1967). However, Oakley (1964) has also indicated a considerably earlier possible age, based
on correlation of the site with the high sea level
preceding the Würm glacial phase altogether.

Another population, and an important one
(though still only partially described—Vallois and
Vandermeersch, 1972; Vandermeersch, 1972),
comes from Jebel Qafza in Palestine. The associated culture is early Levalloiso-Mousterian, suggesting a date early in the Würm (earlier than
Ighoud?). The population shows a preponderance
of modern traits.[22] The best preserved skull has
rather large and inflated brow ridges; in other
specimens these are smaller but still robust structures which nevertheless have stronger marks of
modernity as to which parts of the brow ridge are
emphasized. They also have ample internal sinuses. The braincase is of modern size. In the
best preserved (Qafza VI), again, the forehead is
well developed. At the back, the profile is modern,
though the torus, while placed well down, is on the
heavy side. The face, from the evidence of different specimens, is nonprojecting; facial details are
decidedly modern, more completely in some cases
than others. The lower jaw appears also to be of

---

22 I thank Professor H-V. Vallois for letting me examine and remark
on some of the material, and Professor J. Piveteau and Dr. B. Vandermeersch for showing me the latest finds.

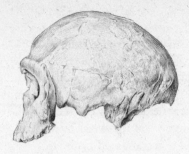

Jebel Ighoud

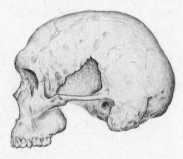

Jebel Qafza

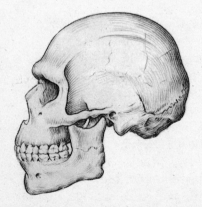

Skhul

Skulls from Jebel Ighoud, Jebel Qafza, and Skhul. (Ighoud and Qafza are drawn from photographs by the author.)

modern form, though the teeth are moderately
large. Vandermeersch (1972) finds the Qafza people in fact to be equally modern in status with the
men of the early Upper Paleolithic, a state I would
argue to be fully, not partly modern. Qafza VI
may be the least "modern" in some aspects, and I
would rate him as archaic in certain details. As
bearing on modernity, here are "generalized distance" figures (based on a number of separate
measurements) given by Suzuki (1970) for different
crania, the distances being computed from an
Upper Paleolithic skull from Afalou, North Africa,
rugged but anatomically modern:

Qafza VI	1.61
Skhul V	2.17 (Mt. Carmel, see below)
Amud	4.49 (Near Eastern Neanderthal)
Shandidar I	4.31 (Near Eastern Neanderthal)
La Chapelle	6.14 (European Neanderthal)
Ighoud I	7.08

The difference from Neanderthals, and the contrast with Ighoud, are notable.
  Finally there comes the better-known Skhul
cave population, from Mount Carmel near the
coast of Israel (McCown and Keith, 1939). It has a
provisional direct date of about 36,000 B.C. (made
on bone of individual Skhul VI by J. L. Bada using
amino acid racemization; I thank Dr. Bada for the
use of this estimate). The stonework is Levalloiso-
Mousterian of the Middle Paleolithic, and the time
has been believed to be late (Higgs, 1961) on other
grounds, probably the beginning of the major warm
interstadial of the Würm, just after 40,000 years
ago. The somewhat varied skulls are essentially
modern, though archaic in their robustness,

strong brows, and a few other traits. The skeleton, which is well represented, is also robustly
modern. The best skull, Skhul V, though quite
modern in its braincase, except for projecting
brows, has large teeth and a forwardly placed tooth
row (like the Neanderthals; see below); but it also
has a good chin. The skulls appear less modern
in detail than the earlier Qafza people, but much
more so than the Ighoud skulls.

These three groups differ in specific ways. A
particular example is the frontal sinuses, which
are generous in Qafza men, and extraordinarily
small in Skhul for such large brows, being missing
entirely in some cases. (Climatic adaptation for
this interesting kind of difference is a dubious explanation, certainly here.) The Skhul people must
be diagnosed (Stewart, 1970; Howells, 1970) as
archaic modern man, with a small number of
Neanderthal-like details which are not general but
distributed over different parts of different individuals. The Skhul folk have often been interpreted
as a hybridizing of Neanderthal and modern men
(or else as transitional). A more apt evaluation,
I believe, would be a modest Neanderthal admixture to an essentially modern people.

It seems clear, in spite of patchy evidence,
that in Europe the progress from erectus to sapiens
began during the Mindel, or second, glaciation,
perhaps 400,000 years ago, perhaps less; and that
approaches to our own skeletal type had been made
by the beginning of the fourth (Würm) glaciation,
and certainly during it. But now comes a puzzling
aspect of hominid evolution: the evidence of Neanderthal populations which overlapped in time those
I have just described, but which were definitely not

modern man, though they must be called <u>Homo</u>
<u>sapiens</u> on the basis of brain development. Recog-
nizable Neanderthals, we saw, are known to come
from the third interglacial but most, as known, belong to the last glacial, i.e., within perhaps the
last 80,000 years.

<u>The Neanderthals</u>. The many finds of the Neander-
thal people stretch from Spain to Iraq and Soviet
Central Asia. A very few of them stem from the
third interglacial: two skulls from Saccopastore
outside Rome, two broken jaws and a broken skull
from near Weimar in Germany, and just possibly
the Gibraltar skull found in 1848. Excepting those
already noted, and possibly one or two others, numerous skulls and skeletons—some being burials,
some ceremonial or cannibal depositions—are believed to belong to the Würm glacial phase and fall,
as far as dated, between 70,000 and 40,000 years
ago, down perhaps to 35,000 B.C. in Western
Europe.

   Though specimens still come to light, it is an
unfortunate fact that their obvious sepulchres in
France, Belgium, and Germany, cave shelters,
were dug out long ago, when archaeology was young
and unaware of the wealth of information which
might have been recovered. The Neanderthal man
himself (1856) was pitched down a cliffside by work-
men, and nobody knows how much of him may have
been lost. Still, the amount of evidence, in skulls
and bones, is impressive; we may even have a few
Neanderthal footprints, preserved in wet mud inside caves.

   Early anthropologists often, and popular
writers almost to a man, saw the Neanderthals as

apelike, half-bent, dwarfish brutes, despicable
Nibelungs unfit to stand beside that child of the
gods, Seigfried (read Cro Magnon Man, the supposedly tall, clean-limbed painter of cave walls).
This was grossly overdone. So was the later attempt to rescue Neanderthal man and make him no
more than a race of modern man, though such
would be a far smaller error. He was unquestionably more distinct than that. This is not invidious.
His fashioning of stone tools, belonging, like those
of the populations described above (i.e., Qafza
and Skhul, who were morphologically more modern), to the general Mousterian tradition, was
skillful and a high-water mark of the early Würm;
and latest information indicates that, perhaps by
exchanging ideas at the end with men of the Upper
Paleolithic (his immediate successors, of modern
European physical type), he produced tools of the
same level of skill, a very high skill indeed.[23]
Certainly his brain size was, on the average, at
least that of modern man (and, according to the
most recent study—Heim, 1970—slightly larger),
with a number of particularly large examples, and
one, the latest to be found, the Amud man of Palestine, having the very high figure of 1740 or more
cubic centimeters.

The Neanderthals varied individually and by
region (i.e., racially). There is, however, a
characteristic morphology.[24] Though capacious,

---

[23] Nevertheless Mousterian work continued to produce the basic
flakes, before retouching, by different techniques from the Upper
Paleolithic people, a diagnostic difference.

[24] This is not meant to imply an absolute uniformity; not all the
traits described occur in all the specimens, any more than in single
modern populations.

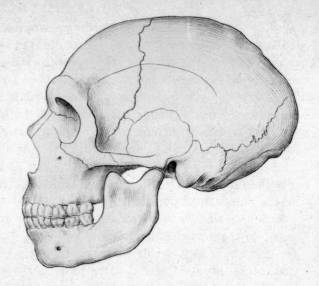

Neanderthal

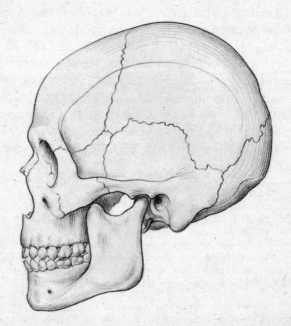

Modern

SKULLS OF NEANDERTHAL AND MODERN MEN

The Neanderthal skull follows a restoration drawing by M. Boule of the skull of La Chapelle-aux-Saints; most of the teeth had been lost during life in the original specimen.

the skull, is long, low, and bulging at the sides; and in the so-called "classic" form of western Europe it protrudes rather pointedly in the rear, although the occipital torus, with the limit of the neck muscles, is placed fairly well down—there is no angulation just here recalling Homo erectus or the Petralona or Solo skulls (see below). In front, the forehead is low though definite, and carries projecting brow ridges of a characteristic kind: fairly even in size all across, arching over each eye, and bending back somewhat at the sides. These brow ridges internally have generous, symmetrical sinuses across the middle part, which, however, never extend upward inside the forehead, above the protruding part of the brow—a difference from many other human groups, modern or early (Vlček, 1967).

But the striking feature of the Neanderthals is the face, which is both high and prominent, in ways unlike Homo erectus. The prominence is not a

projection of the mouth region alone, though in absolute terms the mouth and teeth are set well forward relative to the body of the jaw (a trait of
western erectus jaws which I mentioned earlier
in connection with the differences noted by Stewart.)
For example, the last lower molar tooth is generally placed clearly forward of the forward edge of
the ascending branch of the jaw when the whole is
viewed in profile; in modern jaws the tooth is generally half hidden behind this border, as it is in
jaws of Homo erectus. Thus the Neanderthal mandible appears to have a receding chin, which is
partly deceptive. Actually the upper gum below the
nose is vertical, and the whole profile of the face
is rather straight, because the upper face and nose
are similarly set well forward. The nose, from
what little is known, was probably quite prominent,
and broad as well. The cheek arches slope sharply
back toward the ear, instead of making an angle as
they do in the more retracted faces of all other
known kinds of man (except perhaps Rhodesian
man), who also do not match the long distance from
the ear opening forward to the upper gum or the
profile of the upper face. Thus, in Neanderthals,
the sides of the face, including the outer rims of
the eye sockets, give an impression of being pulled
back, actually due to the extraordinary prominence
of the middle of the face from top to bottom. This
character distinguishes them from such earlier or
contemporaneous skulls as Steinheim, Omo (see
below), Ighoud, or Qafza. The height of the face
also lends a roundness to the eye sockets which is
seen in many Neanderthal faces.

There are a number of special anatomical details around the ear opening (see especially Vallois,

1969) which are common in Neanderthals, though
not all specimens have been examined. In particular the mastoid process, behind the ear, is strik-
ingly puny for such large skulls in the European
examples (it is large in the Amud skull, which
lacks the pointed posterior). Stewart (1961) has
noted a constant trait: there is a fore-and-aft
ridge, the occipito-mastoid crest, on the base of
the skull where the mastoid part of the temporal
bone joins the occipital, on either side of the foramen magnum, the large opening through which the
spinal cord enters the base of the skull. This
ridge projects downward further than the foramen
magnum or any other part of the skull here, which
it does also in Peking man, Solo man, and Swanscombe man. It does not so reach down in modern
man and certain other important fossils (Rhodesian
man, Skhul V).

As to teeth, one cannot make much of specific
differences between Neanderthals and modern man;
size is the most notable effect. Here again is a
tabulation, for the lower teeth, of the crude crown
"area" (length × breadth) giving figures for Pek-

	Incisor 1	Incisor 2	Canine	Pre- molar 1	Pre- molar 2	Molar 1	Molar 2	Molar 3
Peking	41	48	81	86	92	149	150	131
Neanderthal[25]	45	53	73	72	70	129	135	130
Upper Paleolithic[25]	35	41	56	57	61	125	122	118
U.S. white	32	38	55	53	57	115	108	105

---

[25] From Brace and Mahler (1971).

ing man and for Upper Paleolithic Europeans as well.

Insofar as these are good figures, they indicate that the Neanderthals had more than held their own in size of the incisors compared to Peking man, dropping off constantly closer to modern men in the hinder teeth, a trend which evidently continued through the men of the Upper Paleolithic. Greater size in the Neanderthals results primarily from greater breadth across the tooth (from tongue to lip or cheek) rather than from length along the tooth row. This is especially true in the incisors, which have a basal swelling greater than in the incisors of modern man.

The Neanderthals of the Near East (Tabun, Shanidar, Amud) were probably not so generally short in stature as their European relatives; and their skull vaults were not so low or pointed behind. The essential facial projection and length are, however, equally in evidence. And there is at least one peculiarity noted by Stewart (1960) in the pelvis of four skeletons: a lengthening and thinning of the pubic bone never seen in modern man or found in the few pelvic fragments of the European Neanderthals. In both east and west there is a strong tendency for the outer edge of the shoulder blade to have a groove along its back border in Neanderthals and along its front (rib side) border in recent man (Vallois, 1932). In a summary by Stewart (1962), 28 Neanderthal shoulder blades run about 3 to 1 (imprecise because of youth or indefiniteness in some) in the first direction, while a large number of modern scapulae, of all races, run about 4 to 1 in the opposite direction.

The meaning of the feature is unknown, but Stewart
(1964) thinks that in another aspect (inclination of
the shoulder joint) the Neanderthal scapula is more
"advanced" than in many modern men.

Other than the above, only the hand skeleton of
the Neanderthals has had a general comparative
study (Musgrave, 1971), covering specimens from
France, Belgium, Yugoslavia, and the Crimea.
Multivariate analysis shows a community of characters, in a combination which distinguishes the
hand from those of modern or Upper Paleolithic
men. Contrary to general supposition, the Neanderthal palm was relatively long, not short, with
thumb and fingers which had relatively short first
joints and long outer joints (phalanges). But a
principal character was the wide heads of the meta-
carpals (of the palm), just at the base of the fingers
and thumb; and it is indicated that the fingers could
be separated so as to give the hand a particularly
wide span. Other features show that the general
grasp was powerful, and that the intrinsic small
muscles between the fingers were similarly powerful; also that the opposition of thumb and index
finger was better suited to a "key grip" (the way
you hold a key) and less suited to a direct pinch between the tips of these digits, compared to modern
hands. The statistics used confirm these distinctions between modern and Neanderthal seen on in-
spection, but the meaning of the distinctions themselves is not clear, and it would be unwise to read
meanings like "primitive" into them just now.

For the mass of Neanderthal skeleton bones
no such systematic or recent work has been done.
Particular aspects have been noted in the past, especially thickness of bone or large joints. However,

the thickness, both in general or of the walls of
the shafts, of long bones, is variable, and even
the supposedly short forearms and legs may not be
so constant, although the curvature of the radius
of the lower arm is said to be especially common.
It is safe to say only that many individuals must
have diverged in things like bone robustness, especially of the joints, and in minor details, from
what might be found in modern skeletons, again
without suggesting that these were primitive de-
partures. In any case, the old notion of a stooped
neck, and limbs permanently bent at the knee
(which would quickly lead to prostrating exhaus-
tion), has been put to rest.

The Neanderthal problem. Here is a point of importance. What do we mean by Neanderthal? I
have used the term for all the known fossils of
Europe which can be associated with the Würm glaciation down to and into the major interstadial,
which saw their disappearance, i.e., before about
35,000 B.C., as well as some, but not all, fossils
of the Near East in the period (Tabun, Shanidar,
Amud) and of central Asia (Teshik-Tash). This is
a definable time and space as far as Europe goes,
and we may consider that we are dealing with a continuous population. All these specimens partake
of the morphological nature I have described, with
natural variation of individuals as well as some
differences between Europe and the Near East.
And it is a rather striking fact that new finds of the
most recent years in Europe, though some are
juveniles and most are fragmentary, are easily
recognizable as sharing the same character.

There are other specimens—third interglacial examples from Europe, and some from North Africa (two fragmentary jaws from Haua Fteah in Cyrenaica, not very satisfactory)—which have to be viewed as probable members. But I would exclude other bordering populations of the same time zone (Ighoud, Qafza, Shkul) which do not share the specific character I have described.

This is an intentionally confining definition of Neanderthal, for which the scientific reason is the necessity of delimiting a population when appropriate indications occur, as here, in order to allow fruitful comparison with others. The "Neanderthal problem" is as much a problem of anthropologists as of Neanderthals, so let us consider the former for a moment. Many of them have used "Neanderthal" much more inclusively, or else "Neanderthaloid" or another variant, to cover a whole series of fossils running in time from Steinheim and Swanscombe to the Skhul population of Mount Carmel or, in space, forms from every part of the Old World which were not clearly both modern in shape and post-35,000 B.C. in date.

Probably much of this usage stems from the promulgation by Hrdlička (1927) and Weidenreich (1928) of the concept of a "Neanderthal phase" of human evolution. Of course there was a "phase," or morphological zone, of this general kind, to be crossed between Homo erectus and living man. However, to label it "Neanderthal" is to discourage attempts to examine distinctions among late Pleistocene populations, distinctions which might be of great importance, while at the same time to suggest that the Neanderthals, as I have defined them and as they are traditionally thought of, are the

specific source of modern man. We simply do not
know that all men of this morphological zone or
phase were like Neanderthal man proper, or had
his features, particularly the long and prominent
face. Perhaps they were, and some argue strongly
that this is so. Others are strongly opposed. But
what I am pointing out here is that the common
practice of using "Neanderthaloid" to name and
classify, rather than simply to describe, such
morphologically non-Neanderthal people as Qafza,
or others shortly to be reviewed, is carelessly to
state a conclusion in the terminology itself.

Upper Paleolithic successors. This is another
facet of the problem: the manner in which physically modern men, making tools produced from
slender flake-blades, replaced the Neanderthals in
western Europe from about 35,000 B.C. on. The
question is clouded, not only by things just described but also by the fact that the bones were recovered long ago, and crucial evidence of their
tool connections was missed. The Cro Magnon
skeletons, whose name became a general one for
much of this population, were dug out in 1868, and
these were by no means the first. Others, like the
finds at Combe-Capelle, or Lautsch (Mladeč) in
Moravia, might be especially early in the Upper
Paleolithic sequence, and therefore very significant, but the excavation reports are inconclusive
and the sites are spoiled. The Grimaldi "Negroids"
(victims in part of poor reconstruction) and the
Chancelade "Eskimo," though not now taken at face
value, also served to confuse the scene. Although
a large number of Upper Paleolithic skulls and
skeletons in good condition are known, the great

bulk were found more than 40 years ago—most of
them in fact during the nineteenth century, and
thus before the refinement of archaeology and the
various methods of absolute dating such as radiocarbon.

The Neanderthal specimens on the other side
of the divide, in the time of the late Mousterian
cultures, are of course under a similar cloud. So
the zone of transition and replacement remains
hard to specify. The archaeological picture, however (which I shall only touch on), has been under-
going redrafting. The earliest recognized western
Upper Paleolithic culture is Perigordian I, and
most students agree that it shows a combination of
Mousterian and Upper Paleolithic ideas in stone-
working. Therefore it is the archaeological exhibit for the idea of simple transition, and of
Neanderthal-to-modern physical evolution. (It has
unfortunately yielded no human remains.) Its earliest dates are about 33,000 B.C. (Movius, 1972;
Klein, n.d.); its derivation is unknown. It was sup-
planted in many parts (as early as 32,000 B.C. at
the Abri Pataud in the Dordogne) by the Aurignacian, a nontransitional, fully Upper Paleolithic
culture marked by an abundance of decorative objects and other features. The Aurignacian is evidently an intruder, not a descendant of Perigordian
I, and its antecedents are now placed in southeast-
ern Europe (especially Hungary) with an age of
more than 40,000 B.C. (see Klein, in press). No
skeletal remains are known from that time and
place either, and so direct evidence of a modern
population invading the west does not exist. But
the archaeologists personally involved in analysis
of all this assert that "mixed" or intermediate

cultures (such as Perigordian I) do not necessarily signify an actual continuous development. That is to say, while the Perigordian may have had Neanderthal/Mousterian contributors and/or users, it does nothing to deny an earlier existence of the Upper Paleolithic Aurignacian, probably in the hands of fully modern men, to the east.

When Upper Paleolithic people become known in the west (in the skeletons referred to) they are modern in form, not "mixed." They had on the whole rugged skulls with, as we saw, teeth slightly larger than those of modern Europeans (page 99). Such traits have been taken as evidence either of direct descent from Neanderthals or of hybridization with Neanderthals. Though it is entirely likely that Neanderthal genes survive, general aspects of facial and cranial shape fail completely to sustain either idea (Howells, 1970), as does a series of anatomical details (a matter already covered). In fact, multivariate analysis of measurements makes a strong distinction between Neanderthal and Upper Paleolithic skulls, a stronger contrast than between Neanderthals and modern crania.

On the average, Europeans of the present have more gracile skulls, and perhaps skeletons, compared to those of the Upper Paleolithic, but that is the extent of the difference. Upper Paleolithic men continued through the rest of the Pleistocene (at least 20,000 years) in the same form with local variation, and Mesolithic people of the immediate post-Pleistocene maintained the same cranial robustness or a slightly diminished degree.[26] The

---

[26] It is a defect of the hypothesis of direct Neanderthal-modern evolution, in a few thousand years in Europe, that preceding Neanderthals likewise seem to have shown no signs of evolution leading up to

same held for North Africans. Gracility may have
been partly introduced later by farmers coming out
of the Near East, and partly evolved by local
populations on new diets with new cultures; but robustness continued to mark populations of Europe's
northern fringe (Coon, 1939). A different development, however, led to the small-framed Lapps.

The basis of Neanderthal face form. Adaptation to
exposure in regions near the glacier has been put
forward as the selective factor behind the accentu-
ation of face and skull features of the Würm Neanderthals in the west. Here, Howell proposed
(1952), populations of Neanderthals were isolated,
by ice sheets in North Europe and in the Alps and
by unfavorable tundra, from other populations in
the east, except for a corridor along the Mediterranean coast, which then lay farther south because
of lower sea levels. Western Europe, especially
southern France, would have had a windy, snowy
(and generally damp?) climate, though with good
hunting compared to areas to the north and east.
Howell did not, however, go into the mechanisms
by which Neanderthal anatomy actually became
adapted to cold. Coon (1962), pointing out that the
nose was probably prominent as well as broad in
Neanderthals, cited the need for moistening in-
haled air to protect the lungs from cold by means

---

such a change. Brace (1964) has argued that the abruptness of the
alleged evolution was the result of the rapid appearance of the improved stone tools of the Upper Paleolithic, which freed the anterior
teeth from a supposed constant use as a vise in the working of wood
or skin. This led to a rapid diminution of those teeth and thus of the
large and projecting "Middle Pleistocene" face. Brace is one of those
who tend to lump all fossil men preceding the Upper Paleolithic time
period as "Neanderthal." Some others have been unable to perceive
the evidence he uses, that of severe wear on Neanderthal front teeth.

of a large area of mucous membrane, a commonly accepted explanation for racial differences in nasal form. But Coon also suggested that the essential forward projection of nose and face (with a resulting size increase in sinuses of the cheekbone) served primarily to put greater distance between the exposed face and nasal passage, on the one hand, and arteries carrying blood to the brain, on the other, since the brain is very sensitive to changes in temperature.

Support for these general ideas has been produced by Steegmann (1972) in experiments on living young males in two groups, of north European and of Japanese ancestry respectively, with normally varying face shapes within each race. Clothing them comfortably from the neck down, he subjected their heads and faces to drafts of air at the freezing point for 70 minutes (pneumonia, anybody?). He recorded during this period the fall in temperature of the skin surface at several points: forehead, cheekbone, cheek, nose tip and chin tip. Features of shape which appeared best to inhibit drops in surface temperature differed between the two groups. However, indications were that faces prominent in the midline and relatively long, and above all with relatively slanting, non-prominent cheekbones, were best adapted to resisting frostbite; a large head and/or a long one was also important (see Note 3 on page 162).

These experiments suggest that a modern north European fits the "anti-frostbite" model fairly well, with his long and protruding face and nonprotruding cheekbone. They lend special force to the argument that western Neanderthals of the Würm, with their exaggeration of the same characters (though the whole configuration is different), and their long, low, large heads, reflect special pressure from this kind of selection. The results do not, however, explain the similar, if less pronounced, morphology of Neanderthals of the previous interglacial such as Saccopastore (unless these are taken as survivors of third glacial cold selection), nor the still more similar faces of the Near Eastern Würm specimens (Tabun, Amud) unless these were populations extruded from colder regions, or fugitives from the west via the Mediterranean corridor.

2. Eastern Asia

The Solo men of Java[27] are the only Upper Pleistocene fossils of consequence in eastern Asia.[28] The braincases, or parts thereof, of eleven individuals of various ages (and two shin bones) were excavated from a terrace of the Solo River at Ngandong in central Java. No faces, jaws, or teeth survive, apparently having been removed by human action before deposition, which was random,

---

[27] These important skulls have been too much glossed over by writers, many of whom have called them "Neanderthal" or "Neanderthaloid." Weidenreich, the only student to describe them in detail, also used "Neanderthalian" liberally, but with a wealth of qualification it is easy to overlook. He concluded that he had erred in using this term in his early mentions of the Solo skulls, and that they were more primitive than this word would imply. Coon (1962) also was obliged by his criteria of size and angles to classify these skulls as *Homo erectus*.

[28] The poorly known Mapa skull of Kwangtung province in China is almost certainly of Upper Pleistocene date. It is apparently rather like the Neanderthals in form, and might be an acceptable eastern extension of the Asiatic Neanderthal populations.

the skulls having been found at different points in
the deposit, in any posture. Brain size, averaging
1100 cc according to Weidenreich, was little above
Peking man's, and certainly low for men with such
large heads. Skull walls were thick and heavy,
especially in males. Shape was very similar
among all the specimens; and having a whole small
population makes it clear that this shape was characteristic, not the variation of an individual. It
was quite different from the Neanderthals of the
west. Brows were heavy, but straight across the
forehead, neither arched upward over the eyes nor
bending back at the outer corners, where in fact
they were heaviest; a slight notching existed at the
middle. Large internal sinuses were present just
at the midline. The forehead receded directly
from this bar, not rising as in Neanderthals or
Peking man. At the back, a heavy crest, entirely
unlike Neanderthal man's, overhunt the nuchal
(neck muscle) area, which was flat. Forward, at
the long foramen magnum, the base of the skull
was somewhat bent in the fore-and-aft direction.
The occipital torus carried to the ear region and
then down onto the skull's base; neck muscles
must have been massive. Mastoid processes were
large (they were small in European Neanderthals).
The base and ear regions were marked by detailed
features (e.g., formation of the base of the styloid
process and of the foramen ovale, and lack of a
postglenoid tubercle) which are unique, not just
different from Neanderthals. Nothing is known of
the face, of course, except the very top of the
nasal bones, which apparently were not at all recessed under the brows but had the oddly "Gre-
cian" profile of Peking man.

In any event, it was not with Neanderthal men that Weidenreich detected connections; rather, he believed Solo man to be a direct descendant of the Middle Pleistocene Homo erectus of the Trinil zone of the same region, Java.

As a problem, succession in this region has not even reached the enigmatic stage of that in the west. This still unorganized archaeology suggests that many parts of Indonesia (from the Philippines to Timor) were reached, over then-existing land bridges, by men who were culturally, and probably physically, no more advanced than the Solo people, and probably akin to them. This may have been during the early Upper Pleistocene, but perhaps much earlier (von Koenigswald and Ghosh, 1973). The evidence consists of crude flaked tools; a different culture accompanies the appearance of modern man. This last event is marked by a skull dated at about 40,000 B.C., found in the Niah Cave, Sarawak, Malaysian Borneo. Australia was reached not long afterwards by similar people: an occupation of the shore of Pleistocene Lake Mungo in New South Wales which began before 30,000 B.C. has yielded a skeleton dated to about 23,500 B.C. (Bowler, Thorne, and Polach, 1972). This and the Niah skull, together with a few others (Wadjak in Java, Keilor near Melbourne), in form suggest recent Tasmanians or Melanesians. A generally similar population, but one having much more primitive and robust features of forehead and face, with very large jaws, is attested by a number of burials recently found around and close to Kow Swamp in northern Victoria, and dated to about 8000 B.C. (Thorne and Macumber, 1972). Certain others of this character may be older.

Thus there existed in the continent two strains of a generally "Australoid" character during the late Pleistocene. The Kow Swamp people strongly suggest Solo man, but they do not actually show his special anatomical traits, nor can they be said actually to fall out of the range of modern man.

The question is therefore whether Solo man is really involved in the ancestry of one or both of these populations, and if so how. Here his dates become interesting. The Solo river terraces are part of the Notopuro beds, which lie above those bearing the Trinil fauna of the Middle Pleistocene, and do not exhibit the tilting to which earth movements have subjected the latter. The contained fauna is Upper Pleistocene. The animals at Ngandong suggest a colder climate than today's. The prevailing feeling, particularly of von Koenigswald, who participated in discovery of the skulls, is that the terrace belongs to the time of the Würm glaciation, making Solo man a contemporary of the Neanderthals. Cultural remains also indicate a late date. But there is as yet no firm basis for a date of, say, 50,000 years, or for twice or three times that. It seems idle to speculate as to whether Solo man and modern man coexisted, or nearly did; dates are a matter for future solution.

3.   Africa below the Sahara

In 1921 a beautifully preserved skull was discovered in mining operations at Broken Hill, Zambia (then Northern Rhodesia), cutting deep in the ground though the shaft of a sloping, filled-up cave whose original shape and direction could be only approximately determined. Some skeleton bones

(not known to be the same individual) and a second
upper jaw were found as well, along with a few
tools and bones of other mammals. More recently,
parts of another skull (Saldanha) were exposed by
wind erosion at a fossil site, Elandsfontein near
Hopefield, far to the south and 50 miles above Cape
Town. Somewhat, but not much, earlier in its presumed date, this restored braincase confirms in
general the nature of the first Rhodesian man find.
A likely further member of the same form is a
lower jaw from the Cave of Hearths in the Makapan
Valley in the Transvaal, and a possible one is a
fragmentary skull from Eyasi, Tanzania, formerly
called "Africanthropus." Once again, we have a
long, low skull with large brows, and a marked
occipital crest which, however, faces more downward than in Solo man. Though the skull is low,
its flattish sides are reminiscent of modern man,
but not of Neanderthals. The very large brows
have large sinuses which, contrasting with both
Neanderthal and Solo, in Broken Hill pass well up
into the bone of the forehead. The face has the
great length of the Neanderthals, and nearly the
same character of forward prominence and re-
treating cheekbones. The second upper jaw from
Broken Hill is smaller and more modern, and does
not suggest so prominent a face. Although Rhodesian man is duly classed as "Neanderthaloid" by
many, he differs considerably in conformation and
details. The Broken Hill skull gives a brain volume of 1280 cc, low for a big man; and this, along
with tooth size and features of the profile, caused
Coon (1962) to rate him as borderline Homo erectus. Others, more impressed by details, are inclined to consider him an extremely primitive example of modern man.

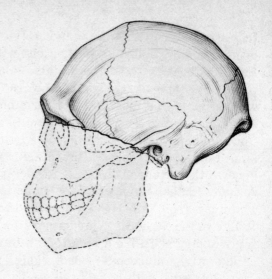

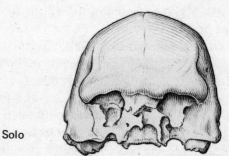

Solo

SKULLS OF BROKEN HILL AND SOLO MEN

The face of the Solo skull follows a restoration as suggested by Weidenreich; no facial parts survived in the original skulls.

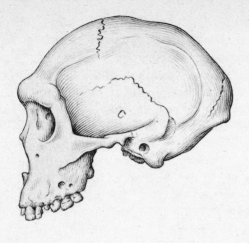

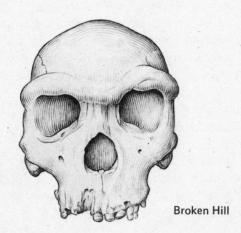

Broken Hill

The Broken Hill skeletal parts look entirely modern, without the obtrusive small deviations of many Neanderthal specimens. Nevertheless, Davis (1964) on close examination did detect some fugitive differences in the positioning of muscle attachments, such as to suggest that, while limb and joint form had attained modern shape, some aspects of musculature differed very slightly. Thus, at the end of a review of the meager remains of the skeletons of early man, we have seen none of whom it can be asserted that they differed in no detectable respect from the normal variation of man today.

Problems of age in southern Africa. For a long time after it was found, chances of getting a date for the Broken Hill skull seemed remote. However, stone tools and animal bones had been found in the cave, and archaeological work in the once marshy area around the cave (by J. D. Clark) gave the tools a stratigraphic background, as the earliest phase of the African Middle Stone Age. The animals were also identified as not older than Upper Pleistocene. The tendency being to equate the Middle Stone Age with the Upper Paleolithic of Europe, Broken Hill man was given an estimated age of 40,000 years (Oakley, 1964). Animals and Acheulean (or post-Acheulean Fauresmith) tools also assigned the Saldanha specimen to the latest part of the Earlier Stone Age, and therefore not much earlier than Broken Hill; the Cave of Hearths jaw was also accompanied by Acheulean tools. The Acheulean has been thought to come down at least to 50,000 years ago. So Rhodesian man became a contemporary of Neanderthal man of Europe, and a rather late one at that.

Here a short digression into African prehis-
tory is important. There has been much able archaeology south of the Sahara in the last 50 years
(i.e., since the find at Broken Hill), with the acceptance of a broad framework of Earlier, Middle,
and Later Stone Ages. The first corresponds to
the Lower Paleolithic of the north, with a long
Acheulean succession of hand axe development.
Beyond this there are no agreed correspondences
with Europe; instead, there appears to have been
continuity of progress, but much local variation in
industrial forms supposedly related to environment—forest versus open country. In the Middle
Stone Age, techniques developed which foreshadow
those of the European Upper Paleolithic; are these
primary or secondary? Absolute chronologies,
though accepting dates for the Middle Stone Age
suggested above (about 40,000 to about 10,000
B.C.), have not heretofore been conclusive.
While there have been many radiocarbon datings,
inconsistencies exist. And correlation of African
pluvial periods with northern glacial phases, once
used as general evidence of dates, is thought today
to be meaningless.

It is now beginning to look as though the whole
sequence has been pegged to wrong dates. A late
Acheulean level has recently had a thorium-
uranium date on bone from Isimila of at least
260,000 years (Howell et al., 1972); another late
Acheulean date of 425,000 years is reported (Isaac,
1972). Earlier dates for Acheulean tools reach
back much farther. This makes much better sense
in relation to the European sequence, though the
end of the Earlier Stone Age could have come later
than the above indicates. However, there has been

a series of newer radiocarbon dates from tool-
bearing stratigraphies; and on the basis of these,
Vogel and Beaumont (1972) suggest that the Middle
Stone Age has been completely misplaced, that it
ended about 35,000 B.C., and that the "full" (not
the early) Middle Stone Age probably began before
100,000 B.C. Geological evidence, in the nature
of Middle Stone Age sites on old beaches in South
Africa, also would relate such sites to the higher
sea levels of the end of the last interglacial (Klein,
1974). So the boundary zone of the late Early and
early Middle Stone Ages would go back a long time,
into the equivalent of the European third interglacial, or even beyond, taking Rhodesian man with
it.[29]

This is not the only important effect. Various
other human remains have in the past been attrib-
uted to the Middle Stone Age, and where such association could be verified the fossils would have
to be dated much earlier than heretofore. Among
them are the Springbok Flats, Border Cave,
Boskop, and other crania. They are either archaic modern or fully modern in form. And such
people might therefore be associated with a Middle

---

[29] At the moment there is much uncertainty on the very important
matter of Rhodesian man's own date. Earlier dating for the Middle
Stone Age, etc., seems to be constantly more strongly supported.
Provisional dates kindly given me by Jeffrey Bada, however, continue
to support a later date for Rhodesian man associations, using bone:
Eyasi, on a hominid fragment itself, about 32,000 B.C.; animal bone
from Broken Hill, about 36,000 B.C.; Saldanha, probably something
over 40,000 B.C. Klein, however (1970 and in press), believes the
stone tools from Broken Hill (Rhodesian Proto-Still Bay) belong to
the end of the Early Stone Age; and from a reexamination of the
fauna and comparison of it with others better known from South
Africa, believes that the Broken Hill and Saldanha skulls should be
assigned to the earliest Upper Pleistocene or the Middle Pleistocene;
that is, that they are 100,000 years old, possibly much more.

Stone Age development of technical ideas which
were important in the Upper Paleolithic of the
northern hemisphere. It must be said, however,
that these finds were mostly made some time ago,
and their credentials are varied, and less reliable
than they might be (Klein, 1970). Some will probably be salvaged. Direct amino acid dating on
bones of an infant from Border Cave gave a result
of the order of 60,000 years, in conformity with
other age indications of the site (Protsch, 1973);
the infant's physical nature is uncertain, but the
modern adult skull is not to be ruled out of association with it. The important Fish Hoek skull (see
below) is now believed by Protsch (1973) to belong
to the latest full Middle Stone Age levels at this
site, granting it a probable age of 35,000 B.C. or
earlier. Richard Klein, however, tells me he
would place its cultural context rather later, in the
earlier part of the Later Stone Age, which would
reduce its antiquity, perhaps considerably. In any
case, this skull is completely modern in form.

Distinctly the most archaic is the Florisbad
skull, which was accompanied by a particularly
early Middle Stone Age culture, the Hagenstadt
variant. Attempts to date the find by various
methods have given results running from not less
than 35,000 years to "over 46,940 B.C." (Vogel
and Beaumont, 1972). Now the Florisbad cranium
could well be viewed as a fairly advanced descendant of Broken Hill. Pushing this last specimen
well back into the past, if the evidence so developed, would remove a longstanding puzzle: how
could so primitive a form as Broken Hill have been
present at the same time as Florisbad, to say
nothing of others still more modern, in Africa and

elsewhere? This applies especially to some finds
farther north in Africa.

Kanjera and Omo. Early in Louis Leakey's scientific career in East Africa he found, in 1932,
fragments of four skulls (unfortunately trampled
into small pieces by Masai cattle before he saw
them) at Kanjera in deposits laid down near Lake
Victoria. The deposits contain hand axes of
Acheulean form and early Upper Pleistocene animals (whose presence implies a date earlier than
that previously accepted for Broken Hill). Oakley
has given a guess date of 60,000 B.C. for the human bones, which may be conservative. The fragments, heavily mineralized, are indubitably from
skulls of modern form, though somewhat low and
thick-walled. In 1967 Richard Leakey, prospect-
ing along the Omo River north of Lake Rudolf in
southwestern Ethiopia, found two skulls of similar
import. The finds came from points a little over
a mile apart and on opposite sides of the river.
Karl Butzer, who did the geology, is satisfied that,
though one was completely exposed by erosion and
the other partially, both derive from the same general level in a long series of deposits 115 meters
deep in all. This is the Kibish formation, laid
down when Lake Rudolph was larger and the Omo
River was flooding higher and higher in this region,
from the end of the Middle Pleistocene until levels
reached their highest point only recently, at the
Pleistocene's end. There are five series of beds
in the whole formation, the lowest, Member I, containing nine such beds in itself. Between two of
these beds there is an interruption, when the water
level fell and the lake edge was occupied by man;
this is believed to be the time of both skeletons.

Such animal bones as occur—not many—suggest
Late Middle or Early Upper Pleistocene. From
several datings of material near its top levels,
Member III of the whole Kibish formation is believed to have completed its formation before
37,000 years ago (Butzer, Brown, and Thurber,
1969). This is far above the skeleton level. Dating by uranium-thorium determinations on oyster
shells from a level just above the skeletons in
Member I gave a possible age of 130,000 years.
The method is considered inexact on such material;
however, it is believed to give minimum ages, not
inflated ones. In any case Isaac (1972) thinks such
a date is not discordant with the general geological
situation. With so high an age, caution is indicated, but an antiquity comparable to that suggested some time ago for the Kanjera pieces is
clearly possible.

The significance of the Omo skeletons lies
partly in this background of dating, and partly in
their good state of preservation. Omo I was a robust and large individual (Day, 1972), otherwise
not different from typical modern man in any particular now known. What is known includes the
braincase, the lower jaw (with a chin of modern
development), and parts of the face, as well as
pieces of representative limb bones. How much he
might have been like the Kanjera people is unknow-
able because of the poor state of the remains of the
latter.

The other find, Omo II, is only a braincase,
also very robust. It approaches modern skulls in
form but is clearly different. It is even longer
than Omo I (215 millimeters, a very high figure)
and is relatively thick. At the back, the area for
neck muscles is rather flat and the crest or torus

across the occiput is well developed, continuing
to just above the ears, and forming in the middle
the most posterior point of the skull, instead of
falling below this point. These things, and the
placement of the greatest skull width immediately
above the ears, are all archaic features. In front,
the brows are well developed but do not form a
bulging, special structure, as for example in
Homo erectus or Neanderthal man. It is surprising that this skull should be coeval with the fully
modern Omo I. And yet the marks of primitive-
ness listed are somewhat restrained, and there is
something of a likeness in style between them.
Omo II must certainly be classed as Homo sapiens
in any simple sapiens-erectus dichotomy, and indeed, allowing for robustness, it must be put
fairly close to modern man. Its backward looks
are, in the opinion of Professor Day (and I agree),
in the direction of Rhodesian man or, especially,
Solo man, or even Vértesszöllös. It represents
an opposite combination of brow and occiput form
from the Qafza or Ighoud skulls.

It is unfortunate that, with all the expert modern work by archaeologists from East Africa
southward, and a fair number of skeletal remains,
the crucial matter of dates still leaves so many
doubts. Have we here in fact an area with an
early occupation or development of modern man?

MODERN MAN AND RACES

The Upper Pleistocene saw the presence in at least
three different parts of the world of men sharing
certain primitive cranial features (Hrdlička's

Neanderthal phase), who were, however, distinct enough in details to argue a real degree of evolutionary divergence among them (Solo, Rhodesian, Neanderthal). The final event was the replacement of all these by fully modern man.[30] We have looked at the problems posed by this in three areas but it needs to be considered as a general phenomenon as well.

Skeletally all modern men are uniform to a marked degree. Racial distinctions have not been established which would approach in degree those noted in the immediately preceding sections. The modern skull is relatively high, with flattish sides and a curved upper border for the temporal bone. Brows vary but only very exceptionally form a full torus; and frontal sinuses are irregular in development, often extending above the level of the external brow ridges (as they do not in Neanderthal skulls). The face is retracted, resulting in a canine fossa, or depression, below the eyes and between the nose and the marked cheekbone angle. The same retraction results in a definite projection of the bony chin at the bottom of the lower jaw, this being less marked in populations with the largest teeth (Africans, Australians). There is other variety in the cranium, which can be specified by measurements for major racial groups: heavy brows in Australians and Melanesians; short faces in these and in Africans; broad skull bases

---

[30] Since *Homo sapiens* is now commonly taken to include the whole post-*erectus* stage, formal terminology generally recognizes modern man as a subspecies, *Homo sapiens sapiens,* distinguished from, for example, *Homo sapiens neanderthalensis.* The name has, however, also been used in Linnaeus' original sense (e.g., see Cambell, 1963), for the European subspecies of modern man alone.

and forwardly prominent sides of the face in Asi-
atics, etc. (Howells, 1973a). The maximum circle
of such variation is not wide enough to include the
skulls of classic Neanderthal or Rhodesian or Solo
men; and the total variation is also less than the
differences between those skulls. We are justified,
on this basis, in viewing living Homo sapiens as a
single polytypic species of no great intraspecific
variety.

The problems are these: (1) How do we explain the special balance of uniformity and diversity
seen in modern man? And (2) how do we interpret
in general the nature of his succession to the earlier man described?

As discovery gathers speed, skeletal remains
of modern man, in most cases showing recognizable characters of the races we know today, are
found at earlier and earlier dates. First let us
note a few cases which are not racially recognizable. The Omo individuals form one: Omo I looks
more European or Caucasoid than anything else,
but study has not been completed. The Skhul people of Mount Carmel in Israel had a pre-Upper
Paleolithic (Levalloiso-Mousterian) stone culture,
although apparently they did not long precede the
beginning of the Upper Paleolithic in Europe. These
folk (in contrast to the Near Eastern Neanderthals
such as the earlier Tabun woman of the same locality) were essentially modern, with qualifications
noted earlier. Racially they are unassignable.
They have even been guessed to be like Australians
or Tasmanians, and there are no grounds for denial.
The Qafza people of the same region are quite certainly earlier in time, and if anything more modern
than Skhul; they also are not recognizable in terms

of conventional races, though the generally fragmentary state of the skulls may be the main reason
here.

In South Africa, the Florisbad skull with its
low vault and forehead—archaic modern in aspect—
is similarly not obviously assignable to a modern
population, and the same is true for certain fully
modern skulls of the region possibly dating from
the Middle Stone Age, or at least preceding all
known populations here: Springbok Flats, Cape
Flats, Border Cave, for example. Like Florisbad,
some of these have been called "Australoid," but
this can be taken as no more than an attempt at
description (and not a very good one), rather than
a serious affiliation.

Otherwise, however, there is considerable
evidence of the early arrival or existence of modern populations in their present regions. The Fish
Hoek skull from the Cape of Good Hope, of Middle
Stone Age associations, is recognizable as related
to the recent Bush-Hottentot population in spite of
its large size and early probable date (Howells,
1973a). In Southeast Asia the Niah Cave skull has
been diagnosed by Brothwell (1960) as a youth of
general Melanesian-Tasmanian form; the population represented by this skull must have occupied
Indonesia before the arrival of Mongoloid peoples
from the north (Howells, 1973b). This population
had spread to Australia before 30,000 B.C., since
the Lake Mungo skull gives every sign of belonging
to it.

For eastern Asia and India evidence is slight
or absent. It is partly negative for America,
where entry of man from Asia should now be reck-
oned at about 25,000 B.C. if not earlier (McNeish,

1971). No skeletons so old are known, but later
ones from the end of the Pleistocene or after
(Minnesota, Tepexpan, Midlands) give no indica-
tion of being anything but general American Indian
in form, suggesting that firstcomers were much
like the last, up to the time when the Bering Strait
bridge flooded, leaving only the Eskimo-Aleut im-
migration to follow.

We have reviewed the much fuller record for
Europe. The men who everywhere replaced the
Neanderthals, about 35,000 B.C., were not only
modern, they were recognizably Caucasoid. But
difficulties in explaining the European picture in
particular, and that of the Old World in general,
remain. The greatest is this: whence came the
populations of modern man, in some places already
distinguishable as forebears of the living people of
the same region, as in Europe or the Southwest
Pacific? No single hypothesis can be upheld ex-
clusively, and controversy goes on. Solutions fall
into three types.

One very general explanation of the whole
transition is Weidenreich's version (1946) of the
"Neanderthal phase," brought up to date by Coon
(1962). In this, populations of Homo erectus, differing in Africa, or North or South Asia, progressed along similar lines, all reaching a stage
of modern man but continuing to exhibit "racial"
differences. Thus Peking man gave rise to Mongoloids (and American Indians), and Java man led
to Solo man and thus to the Australo-Melanesians.
Neither Weidenreich nor Coon took the western
Neanderthals as parental to the Europeans, however, accepting instead that they became extinct
on the arrival of the Upper Paleolithic tribes.

Another view is a revival of Hrdlička's (Brace, 1964; Brose and Wolpoff, 1971), that there was no such extinction, and that the Neanderthals evolved directly into modern man. The key change is taken to be a reduction of the face and teeth consequent on making better stone tools and so abandoning the supposed use of the front teeth as a vise to hold materials being worked on with earlier tools of less efficiency. (In this view the meaning of "Neanderthal" is the expanded one described earlier, so that other populations such as Solo or Rhodesian man are included as parental populations, as Weidenreich held, though the writers are not very specific.)

A final view (e.g., as expressed by Weiner, 1958) is more flexible. There was general Middle-Upper Pleistocene progress, and a degree of divergence among geographically isolated populations, leading to such forms as Neanderthal, Rhodesian, Solo, and modern man, and possibly others. By this zoologically natural pattern, modern man could have descended from one of the above, or independently, but not from all.[31] Such a population, eventually spreading out, partly replaced other populations, and partly preserved these others only in the genetic sense, by receiving some admixture from them. A difficulty with this explanation is that the original home of such a population is not known, unlike that of the Neanderthals of Europe and western Asia. The gathering evidence, however, would place the examples of

---

[31] According to Birdsell (1967), Weidenreich, shortly before he died in 1948, shifted his ideas and in conversation stated his belief that the Australian aboriginals probably did not descend directly from Solo man of Java, but instead were a population of modern men which was affected by Solo admixture.

modern men who precede the Upper Paleolithic in-
vaders of Europe in a general southern zone, from
South and East Africa through the Near East to
Southeast Asia. India, unfortunately, is a blank,
little being known either archaeologically or skeletally.

The same evidence seems to reinforce the
general view that Neanderthal man (in the strict
sense), far from serving as the ancestor of recent
man generally or the Europeans in particular, became increasingly isolated as modern man was developing and expanding elsewhere during the late
third interglacial and early Würm glacial. Thus
the seemingly catastrophic replacement was a real
phenomenon, a contrast between two forms of late
Pleistocene man who had diverged over some
time.[32] For the Neanderthals indeed had a more
individual morphology as we have reviewed it, and
one seemingly farther off the path we might recon-
struct as leading simply from Homo erectus to
living man when compared to the Rhodesian or Solo
men. I have favored the Coon-Steegmann explanation of cold selection for the Neanderthal face over
Brace's supposition of residual large size and of
use of the teeth as a third hand. But it is fruitless
just now to be positive, with such poor information
on the other Upper Pleistocene men. While we can
make increasingly good assessments of the differences among the fossil men, we need better explanations for the differences.

---

[32] This is not to say that all Neanderthal populations everywhere
came to the same sad end. Thoma (1964) finds evidence from measurements and general form to derive certain central Asiatic Mongoloids from them. Personally, however, I regard all populations of
modern man as descended from one original main source, however
hybridized with other surviving forms some of them may have become eventually.

The Pattern of Evolution in *Homo*

MORE ON CAUSES

I have reviewed the physical and fossil evidence of
the course of hominid evolution. A body of theory
is developing around function, ecology, and behavior, those things which do not fossilize but without
which the evolutionary separation of man from
other primates is incomprehensible. Progress in
this depends on what inferences can justifiably be
drawn from archaeological and anatomical evidence
and from the study of living primates.

We have considered why hominids diverged
from pongids (diet) and how australopithecines became adapted to the ground and open country (bipedalism) and began the expansion of cerebral control of manipulation (with the making of stone
tools). The obvious changes in Homo erectus were
larger body size than his ancestor, Australopithecus, and a dentition probably adapted in its proportions to a diet with more meat in it. And we
shall now see evidence that Homo had become a

hunter of big mammals. It is perhaps no coinci-
dence that this period saw the extinction of the pre-
sumptively vegetarian Paranthropus, as well as of
the large ground ape Gigantopithecus, which had
been present in Asia for well over five million
years. Since, on the evidence of their earlier co-
existence with Australopithecus, these two animals
were, as species, not in serious competition with
the latter for food and space, and since they had
clearly survived nonhominid predators on the
ground, the inference is that they unexpectedly
qualified as big game animals for the newly
emerged Homo. Here was a new carnivore from
whom they could not escape. For he was a cousin,
much like them in gait and perception, but one who
had now evolved a strong appetite for meat, a de-
termination in hunting, a weaponry, and a degree
of social cooperation.

HUNTING

The meaning of a hunting existence, as molding
human development from this point down to the end
of the Pleistocene—many millennia after the appearance of modern man—has been much stressed
recently (see, e.g., a summary by Washburn and
Lancaster, 1968). Its probable role in social organization, sexual distinctions, and division of
labor, and the acceptance of killing as pleasurable
and aggression as natural, have been pointed to.
So have its likely effects on brain development,
through stimulation and the demands of alertness
and cooperation, and through the supply of protein
afforded by meat. Direct evidence is slight, but
it is not lacking, and in any case we can reorganize

the outlines of what must have taken place.

Overall, it is obvious that there has been a shift from the basically vegetarian diet of pongid hominoids to that of man, in which animal protein is of signal importance. But we must qualify the terminals in the transition, that is to say, the view of vegetation versus meat as the diet. Primates are certainly opportunist eaters of meat. Baboons eat young gazelles they happen upon (DeVore and Washburn, 1963); gibbons eat eggs and nestlings. Chimpanzees hunt, kill, and eat small or young mammals—monkeys, antelopes, and even infant chimpanzees (Teleki, 1973; Bygott, 1972). The associated behavior, in chimpanzees, includes marked excitement and interest, but also indicates that such killing and eating results from chance encounter—that is, they do not start a hunt unless prey is seen or heard—and suggests it is less deliberate than the termite-fishing discovered by Goodall, already mentioned. Nor does the manner of eating the meat obtained argue a dietary need, or modify the idea of fruit and similar vegetable materials (Nissen, 1931) as the basic diet, in adaptation of the animal or in the bulk consumed. In addition, chimpanzees and gorillas are specific in this respect, the former subsisting largely on the reproductive parts of plants (fruit, seed pods) and the latter on material of the plant itself (wild celery, inner bark, roots).

At the other terminal it is obvious that, toward the end of the Pleistocene, man in Europe and America was the killer and consumer of mammals which were relatively both large and numerous. This situation was surely a late and special development. And it is a fair complaint that man the

hunter has been extolled at the expense of woman
the gatherer. Archaeologically, signs in living
sites of vegetable food are furtive compared to
meat bones. Nevertheless ethnographic studies
make it clear that, in temperate or tropical regions, "hunting" peoples depend heavily on gathered foods. Part of this is expressed in the classic division of labor between sexes: women,
restricted (by the attention they must give young
children) to within the radius of a fraction of a day's
walk from camp, gather what food can be gathered.
Richard Lee, in fact, has been quoted as suggesting that a carrying container (basket, bag, or
pitchi, of string, skin, or wood) should be looked
on as probably the first hominid invention, of great
importance in bringing back more than hands can
hold. (Try picking string beans or raspberries in
your own garden without a basket.) And this is
only part of gathering. Men, spreading out on be-
half of the whole economic group after large animals (which they may fail to find or kill), will also
gather up whatever else they encounter, to take
home if they can or to sustain themselves in the
field. Thus the possibilities for food do not con-
sist simply of veldt cabbage or wild yams on the
one side and elands or kangaroos on the other, but
run a gamut through berries, gum, fruit, honey,
honey ants, grubs, eggs, nestlings, and "slow"
game of any kind: snakes, lizards, or mammals
such as opossums or porcupines.[33] (All this

---

[33] The Tasadays of Mindanao, lately rushed into public notice,
seem to have subsisted as pure gatherers, eating for animal food only
such things as tadpoles, grubs, crabs, and small hand-caught fish. I
suspect, however, that their sensational innocence is a manifestation
more of cultural backsliding than of a primordial state.

ignores shore foods, which probably played no measurable role in hominid evolution.)

Such is the range of accomplishment within which we should view man's increasing ability to benefit from animal protein through the outright, culturally assisted and mediated finding and killing of mammals too large and too fast for single early hominids to overcome. It presupposes increased size and strength in the hominids themselves, tools for killing and butchering, and a social structure which would allow males to hunt more freely in groups while females foraged to fill the possible hollows between successful kills—the steadying background of more constantly available foods.[34]

All this is hypothetical. What can be said about actual steps along the way? Can we observe the replacement of opportunistic hunting and scavenging by the cooperative search for large game, to be butchered, brought back to a known meeting place, and institutionally shared? Isaac (1972) has reviewed archaeological evidence, principally for East Africa, of such things in the Lower and Middle Pleistocene.

At Olduvai, intensive excavation of entire local floors by Mary Leakey has shown some camping or "home" sites, where stone tools and food bones are mingled, as well as places where it is evident that one large animal was butchered. These go back nearly two million years. One site,

---

[34] A possible preadaptation to human ranging for food is the behavior of the large apes, whose groups move more freely and widely compared to gibbons or monkeys, and whose social units are looser. Groups may come in contact, and more readily receive or release individual members, so that while such groups are real entities they also do not necessarily engender dangerous hostility in encounters.

in the Koobi Fora formation of the East Rudolf region, with the bones of a single hippopotamus, may carry such signs back to 2.6 million years, the level of the earliest tools yet known (Isaac, Leakey, and Behrensmeyer, 1971). It cannot be determined, however, if these remains of large animals represent actual kills rather than scavenging of animals already dead: at the other principal tool site of this age the bones are of smaller species. It seems clear that the hunter, whom we presume to have been Australopithecus (or "Homo habilis"), was still of modest size in body and brain. Nonetheless he must have been a fairly industrious tool-maker: one site in Bed II at Olduvai is a workshop, yielding not tools and food remains, but large amounts of flakes produced by the making of tools of chert taken from an outcrop close by.

Sites of Homo, in the Middle Pleistocene of Europe and North Africa, show changes. The tools are Acheulean hand axes (beginning as early as one million years ago—possibly 1.5 million years by one dating), and they are decidedly larger than the earlier Oldowan tools. They are also more skillfully formed, and while they may have been an all-purpose implement, actually used in hunting, they would be most useful, and probably necessary, in the efficient butchering of large animals with skin and sinews in a freshly killed state. In addition, the refuse bones show that the animals eaten were, on the average, of larger size, with concentration on the medium-sized to large antelopes and equids (zebras), rather than on the smaller antelopes and pigs more prominent in Lower Pleistocene leavings. There also appear sites with special concentration on one or two preferred species alone, with

other evidence which for the first time rules out
scavenging as the explanation, in favor of organized surrounds and drives. All this suggests that
Homo was the first full-fledged hunter. Australopithecus had prepared the way; he was certainly in
part a meat-eater, else why the tools? But we
cannot say he was not still mainly accepting the
meat of any animal, dead or alive, with which he
could cope, rather than organizing his hunts.

LANGUAGE

Though not seen in bone or stone, language is such
an absolute in the nature of man that we are obliged to consider its evolution. With it I would as-
sociate mentality—in common terms, facility with
ideas—and also esthetic development, though without stressing the latter. These are the foundations
of human culture, which we may here look on simply as their output. Other fundamentals of our
present species are either out of the scope of this
essay (such as social organization and behavior) or
else so speculative that they get more treatment
just now in popular writing than in scientific (for
example, secondary sex differences—but see treat-
ments by Holloway, 1972, or Campbell, 1966).
But language and mentality are unavoidable.

In viewing them we must understand again that
there is much we can learn from other primates,
and much we cannot. What we can learn is the
wealth of preadaptation for the human condition:
the particular use of the senses and modes of perception of the higher primates; the complexity of
social adjustment and the imperative to be a social
animal; the slow growth and maturation, and long

learning period, which allows such social development in an individual, as well as great flexibility
and large stores of memory in all kinds of response; the manual skill, resting on coordination
of excellent vision, good hands, and large cortical
representation in the brain, a coordination also
permitted by slow maturation. Looking only at
perception, this means than an ape knows far more
about his environment than do other animals. An
object may fall as clearly on the retina of a dog,
who may mouth it or bite it, and is sure to sniff it.
A chimpanzee will handle it, nibble it, and look at
it with his binocular vision; in color, texture,
weight, movability, strength, and so on, it takes
on more properties for him.

In fact, as time passes, students see more
and more "human" characters in this closest relative. For 50 years laboratory work encouraged
workers, in spite of themselves, to "humanize"
chimps. But recently Jane Goodall showed them
making tools (to fish for termites), not simply using tools under laboratory situations. And Kortlandt (1969) found that wild savanna-living groups
of chimps not only walked bipedally more often
than forest groups but also used natural objects for
weapons much more readily and more accurately
than forest-livers, even though the latter did indeed fashion clubs from branches. Kortlandt hy-
pothesizes that man himself has dehumanized the
chimpanzee, by long ago driving him almost completely out of open country (perhaps at the same
time he was putting an end to Paranthropus, who
could less easily take refuge in the woods). So,
he says, we are not seeing all of the chimpanzee,
but only that much of him which can develop in the
forest.

In all the above we see the prehuman condition, surely already in the baggage of Ramapithecus as he set off on his hominid journey. What about language? Here it is essential to recognize the distinguishing nature of human speech as vocal-auditory, and to understand that there were two necessary but separate patterns of development involved in the growth of language. On the one hand, evolution had to produce a sound-making apparatus capable of the complexities of our phonemic system, together with the cerebral basis for encoding and decoding the sounds. On the other, it had to raise other cerebral capacities to the point of being able to generalize from perceived experience, and thus to use vocal symbols for the generalizations (e.g., "yes, no," "then, now") without which specific objects cannot be linguistically related. These two patterns form one phenomenon, and they reinforced each other as they progressed; but they are separate, a fact which is important in understanding their development.

There has been a tendency among anthropologists and others to discount the vocal-anatomical differences between, let us say, chimpanzee and man and to suggest that a chimp's linguistic handicaps lie simply in the cortex of the brain. If only he were capable of the symbolic projection involved in language, utterance would not be a problem; as the Duchess told Alice, "Take care of the sense and the sounds will take care of themselves." But at present it appears more likely that evolution of the perceptual and symbolic processes—the sense (that aspect of language which allows us to convey the same ideas whether the speakers are using English or Bantu or Chinese)—could have proceeded directly from chimpanzee to man. The anatomical evolution—the sounds—now seems, on the other hand, both divergent from pongid developments and very important in the whole scheme. It invites the "motor theory" of speech perception (Lieberman, Crelin, and Klatt, 1972), whereby the decoding of the rapid and subtly precise sounds of human language depends partly on the experience of the muscular articulation needed to produce them. If so, the brain could not have evolved the ability to decode human speech without evolution of the human vocal anatomy as a necessary tool in learning. In turn this argues that elaboration of language capacity acted powerfully on the evolution of everything between the larynx and the lips (as well, of course, as on the brain), in fact eventually producing a mechanism less well adapted for swallowing food than its hominoid predecessor, as it struck a balance between the needs of breathing, eating, and speech.

Let us consider first the aspect of perception, conceptualizing, and symbolizing—the sense. Here, once more, chimpanzees now seem to be revealing more than before. Communication, of course, is highly developed among them as among other primates, by varied vocalizations, by gesture, expression, posture, and touch: a rich repertory which need not be used in rigidly defined patterns in order to be effective. But their relatively lavish vocalizing, not characteristic of orangs or of gorillas, perhaps masks the actual vocal divergence of chimp and man. It cannot be called "speaking." It is not language—it is signals, which deal only with an immediate situation (whether emotionally charged or not). Even if, for example, one chimp solicits another for a purpose which is out of sight, both have to be familiar with the purpose; number one is not equipped, vocally or mentally, to say "Want to see a pink elephant?"

To press matters, however, chimpanzees have several times been introduced to human language. (Obviously, this is no longer a study of chimp communication.) First came futile attempts to make them "talk." In the wild they constantly use a variety of open-mouth calls. The great contrast with man lies in the ready babble of human infants, when beginning to talk, using consonantal stops naturally and easily (da da, ma ma, etc.), as against the gruelling struggles of young chimpanzees to articulate any such sounds, under pressure of all the wheedling and food bribery that remorseless psychologists can bring to bear on them. In separate efforts, each lasting many months, a chimp and an orang learned to speak three words (mama, papa, cup). Even this was accomplished only by manipulating the animal's lips at crucial points, and the result was not clear tones but something between a grunt and a whisper. The attempts were then abandoned since progress did not promise to get easier.

But it has long been realized that this contrast is not the measure of the real differences or likeness in the desire or ability of man and chimp to "communicate," or to manipulate (or rather "mentipulate") symbols cerebrally. This has been shown dramatically in recent work. The Gardners (1969) substituted standard deaf-and-dumb sign language for sounds produced vocally, and their subject chimp, Washoe, rather rapidly built up an

impressive "vocabulary" of these, showing also
that this was no case of teaching Rover to roll
over. For example, where the Gardners referred
digitally to the refrigerator in standard fashion,
as "cold + box," Washoe invented her own name
for it, much more logical from her point of view:
"food + drink." Another young chimp, Sarah, has
learned to read and write a sort of chimpanzee
Chinese (Premack, 1971, 1972). Sentences,
formed by choosing from a large set of plastic
symbols of different shapes and colors to serve
as ideograms, are put vertically on a magnetic
board, by Sarah or her interlocutor. They are not
great literature (as yet). But such examples as
"If Sarah take banana then Mary no give Sarah
chocolate," or "red and yellow are colors," or
"cracker is not fruit," show that abstractions and
generalizations (negative, same/different, condi-
tional, plurals, classes of objects, color, etc.),
expressed by symbols, are, as long believed, well
within ape capabilities.

This is true linguistic behavior; and Washoe's
feat of making a new combination of learned words
to say "refrigerator" is an example of "produc-
tivity," Hockett's (1960) term for this specific attribute of human language. The qualifications, of
course, are, first, that chimps would not invent
and use such "language" even if their habitat were
stocked with sign boards and, second, that it is
not vocal-auditory. The route to speech was not
through chimpanzees: they have the conceptual
preadaptations but not the vocal. As Premack
says, "Causing objects to adhere to a vertical
surface is something [Sarah] did easily, in contrast, for example, to producing human sound."

There had, instead, to be a considerable separate evolution of vocal anatomy (with, of course, associated cerebral controls and auditory association). This is the other aspect, or prerequisite, and it can be viewed to a small degree through the evidence of fossils. The sounds of human language are vowels, produced in the larynx, modulated as to frequency and overtones in the pharyngeal space (supplemented at will by the nasal cavity) and filtered by consonants formed in and around the oral cavity, the mouth, i.e., by lips, teeth, tongue, and palate (Lieberman, 1968; Lieberman, Crelin, and Klatt, 1972). In modern man, with his upright stance and face retracted under the skull, everything has worked out in favor of this particular system. The oral "box is deep, with a high palate arching over it and a depressed floor lying between the opened-out sides of the short lower jaw; and the teeth can form a closed wall around the tongue, making their combined work efficient in consonants like "l," "th," or "d." The whole tract is sharply bent behind the tongue and the larynx is well down the throat, so that this part of the vocal tract is open and deep. Of major importance is that the back of the tongue extends well down the throat, forming the anterior wall of the upper vocal tract and being effective in modulating the resonances, or formant frequencies, of the vowels. The tongue lies entirely in the mouth in other primates. The large apes are all wrong in this and other respects (Kipp, 1955). The palate is flatter, the anterior teeth slant forward, the jaws are long, and the large canines are associated with a diastema or gap in the tooth row, all inhibiting the production of consonants. Behind,

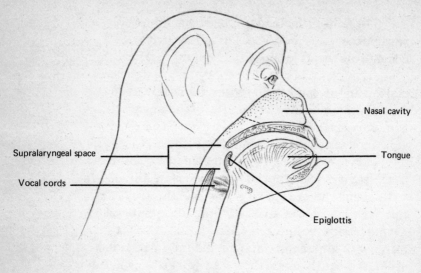

Chimpanzee

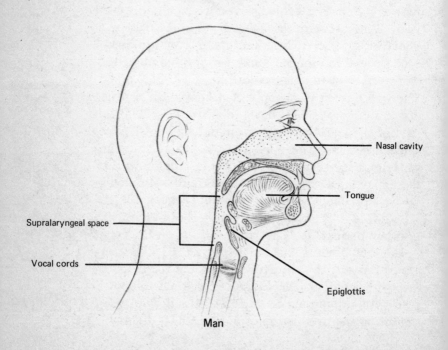

Man

VOCAL ANATOMY IN MAN AND CHIMPANZEE

Principal details of anatomy between vocal cords and lips affecting
the production of speech. Of particular importance (see text) are
the angulation between the mouth and the pharyngeal space, the
greater length in man of the pharyngeal space, and the fact that
the back half of the tongue lies in the mouth in the chimpanzee
but forms the front wall of the long tract above the vocal cords in
man.

the angulation of the vocal tract is less, the tongue
is long and thin, and the larynx is higher up, limit-
ing and interfering with the pharyngeal space. The
big apes also have large larnygeal sacs (smaller
ones exist in monkeys), or air pockets to the side
of the larynx, which apparently contribute to the
volume of sound but diminish the specificity with
which it can be modulated.

These differences obviously explain a major
part of the difficulty met in trying to make chimpanzees "talk," which is a use of voice for which
they are quite unadapted. Kipp, in fact, believes
that the large apes have become rather specialized
in the above features, compared to gibbons or
monkeys, especially in their larger laryngeal sacs.
In this view, they and man's ancestors have been
diverging from a more similar set of structures,
and there may be a hint, in the development of the
laryngeal sacs, that large body size came after the
lineage separation, as the fossils of Ramapithecus
have already suggested.

When did language "begin"? There is no reason to think that Ramapithecus, especially if he

had procumbent front teeth, was any better at human sounds than a chimpanzee. But the trend to face shortening and deepening was apparently set; and the bipedal australopithecines, with closed and vertical tooth rows, had at least the beginnings of the right architecture.[35] In addition, it is argued that upright bipedalism itself was a major factor in angulating and reforming the upper vocal tract, and thus removing handicaps to hominid speech (Hill, 1972). More was needed, however, than the capacity to make human sounds. Some slight original switch had to set a trend toward using such sounds for socially useful communication, ending by exerting a powerful force on further development both of sounds and of the anatomy producing them. I would agree with Holloway (1972) that the original manifestations were probably early, in the Ramapithecus-Australopithecus transition, and first progressed in connection with social affect and control rather than hunting, i.e., where the conversation, so to speak, was general, not confined to a special activity of adult males. Many inputs, however, tool-making and hunting among them, would push the trend, making useful the aspect of "displacement" (Hockett), or the carrying of information about matters displaced from the utterance itself in space or time. We well know that evolution toward and within Homo saw constant brain growth and tooth reduction, always with

---

[35] Lieberman (1968) believes that the Makapan *Australopithecus*, as restored by Dart, was close to a chimp in palate, mandible, and pharyngeal tract, and so could not have produced human speech. Whether this is right, in palate and mandible "Telanthropus" and *Homo habilis* (e.g., OH 24) had apparently steepened the front wall of the former and opened out the latter significantly.

maintenance, and probably improvement, of the
form of the oral cavity for sound modulation, up to
modern man. We can hardly be wrong in supposing
that this has been a strong positive selective factor
in the form of the face throughout.

LANGUAGE, MEMORY,
AND EXPANSION OF THE BRAIN

The demands of modern languages on modern man
are a vocal anatomy, the ability to abstract and
symbolize, a full nervous coordination of the receiving (hearing), decoding (understanding), and
transmitting (speaking) of vocal symbols, plus a
large brain storage and production capacity for
vocabulary and for the niceties of grammar which
distinguish past from present, actual from possible, and so forth. Implied in this is the demand
for a prodigious memory, something to which we
shall return.

Look contrariwise on the demands of modern
man on language. They comprise not only all the
above niceties of grammar but also great specificity of reference in naming and describing, and
an absolute divorce from affective content or emotional state--the most violent statement can be
made in a whisper, or with the flatness of a printed
page. In this medium we meet our needs for social
understanding and convention; in it we store our
cultural tradition (which we can now fossilize in
libraries); in it our children learn and mature.
And it is obviously in this medium, of being able to
frame ideas in clear and specific symbols, to others or to ourselves, that evolution of brain and
mentality proceeded. As a chimpanzee, because

of his ability to see and handle, can know more
about an object than a nonprimate can, so does a
man, because of language, know more about it than
a chimpanzee: he has a name to remember it by,
he knows—or can ask—whether it is near or far,
rare or common, useful or not, part of another
object, and so on. Certainly man is in a different
universe. Though chimps and other higher primates are, for animals, extraordinary at learning,
and even at learning to learn and at remembering
solutions of laboratory problems, there is still a
big component of disorganization, of having to begin all over, in approaching fresh problems. Nor
can a chimp learn from another chimp except by
watching and copying, at which chimps are good.

Human speech is particularly persuasive as a
factor fostering brain increase and development,
a sort of overlord to the other suggested factors of
hunting, tool use, and social ordering. Spoken
language and mental capacity, we may assume,
formed a feedback loop leading to modern man.
That is to say, the more the brain gave depth and
flexibility to speech, the more scope the latter
had in social control, in successful hunting, and
in deploying and teaching technical skills. ("No,
you must put the second knot underneath, or it
won't hold.") Here, of course, is epitomized
again the idea that culture became the milieu in
which natural selection largely operated on Pleistocene hominids to foster mental development.

It is not possible to be much more specific as
to the whys and whats of brain increase. For one
thing the fossil evidence consists not of brains but
of the insides of skulls. For another, precious
little is yet known as to why a big brain is actually

better functionally. We have to fall back on the
simple figures of increasing cranial capacities.
The following brief table gives figures from Tobias
(1971a, 1971b) with a little rounding, and from
Holloway (1973).

Chimpanzee, male	400 cc
Chimpanzee, female	370
Gorilla, male	530
Gorilla, female	450
<u>Paranthropus</u>, male (4)	517 (Holloway, 1973)
<u>Paranthropus</u>, female (ER 732)	506 (Holloway, 1973)
<u>Australopithecus</u> (6) South Africa, male and female	442 (Tobias, 1971a, p. 494)
<u>Homo habilis</u> (4) (male and female?)	637 (590–687) (ER 1470 = 800?)
<u>Homo erectus</u> (6) Java, male and female	860 (750–975) (Pith VIII = 1029?)
<u>Homo erectus</u> (5) Peking, male and female	1043 (915–1225)
Modern male	1450
Modern female	1300

Modern brains vary greatly in size: geniuses
and very ordinary people have had small brains
(the size of Peking man's) as well as large ones;
and microcephals with brains no bigger than an
ape's still may learn to talk and interact in human

fashion (Holloway, 1966). Nonetheless we are big-brained as a species: the average figure has indubitably risen greatly, and the conformation of any modern skull is that of a large braincase fully overhanging the ancient vertebrate sensory, respiratory, and masticatory machinery (i.e., the face). How are we going to think about this?

First, we must remember that, set against the brains of other primates, our brain is a big one but also a different one. This is borrowing Holloway's point that equal cranial capacities are not necessarily measures of the same thing. That is, the slightly larger brain of the australopiths, compared to apes, probably involved reorganization (by shifting interactions of different components, for example, according to Holloway) more than mere neuronal increase. Tool-making suggests this from the empirical side (though at the habiline stage brain size was picking up). The unavoidable conclusion that vocal communication in the australopiths was quite different from that in chimpanzees, and was at least incipiently human language, argues the same thing logically.

We are not going to know whether the cortical speech centers (Geschwind, 1972) had come into being, but any supposition that true speech existed, even incipiently, demands their presence. B. W. Robinson (1972) suggests that two brain systems must be recognized as involved in speech, not one only. The limbic system, comprising deeper and older parts of the brain, controls the typical vocalizations of primates; these brain regions are also involved in emotional behavior, a fact which would explain the strong affective quality of monkey and ape calls. The limbic system, it was

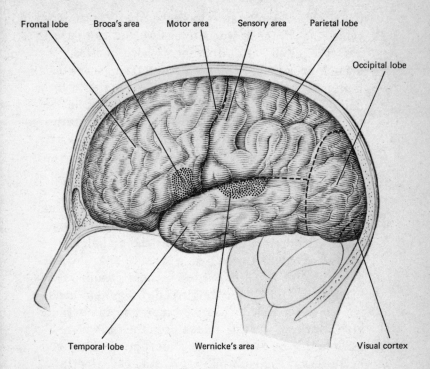

Frontal lobe   Broca's area   Motor area   Sensory area   Parietal lobe

Occipital lobe

Temporal lobe   Wernicke's area   Visual cortex

PARTS OF THE CEREBRAL CORTEX OF THE BRAIN

Shown on the left half of the brain are the main divisions of the cortex (frontal, parietal, temporal, occipital), as well as certain specific function areas (e.g., visual cortex, motor cortex). Broca's area and Wernicke's area are particularly important in articulating and in comprehending spoken language respectively.

found, is evidently able to evoke simple ejacula-
tions (though these were actual words, not calls)
during excited states in a man whose cortical
speech areas had been so traumatized as to render
him speechless in normal emotional states. In
some components of this part of the brain, man
has a size increase over other primates (hardly
substantial in gross terms); in others, man and
monkey are equivalent in size (Holloway, 1966).
The limbic system is also undifferentiated as to
side of the brain.

The other system is the cortical one of "true
speech," largely independent of old-brain emotional connections; it comprises the classical
speech areas of Broca and Wernicke (Geschwind,
1972), which are limited to one hemisphere, normally the left. Their original development need
not have demanded a marked increase in cortical
size, perhaps calling instead for the kind of re-
structuring of pathways and interactions stressed
by Holloway. From this view, an australopith
possessed of rudimentary speech, and a chimpanzee having a totally divergent vocal development,
might, as we have seen, have endocasts (molds of
the inside of the braincase, all that can be recovered from fossils) not discernibly different in
size and shape.

What, then, lies behind the great increase of
cortical bulk and area?[36] This is where answers
have been vague and unsatisfying. Areas known to
be related to specific functions (e.g., the visual or

[36] Much discussion of the matter has been cast in terms of "excess
neurons," i.e., the increase, by billions of nerve cells, in primates and
hominids, of the brain over what might be called the housekeeping
necessities of the species in terms of its body size and general grade
of organization. See Tobias (1971b) for a review and figures.

the motor cortex) have not expanded commensurately. The importance of association areas, surrounding these other centers, or composing much of the frontal lobe, has long been evident. Here, in the simplest terms, it has been assumed that sorting out and evaluating of primary information after it is received by the sensory centers, for reference and action, takes place; and the significance of this to higher mental activity has been taken for granted as an evolutionary force. But perhaps a more specific function makes a better explanation: memory. Rensch and associates (Tobias, 1971b) found that, across different mammal species, those with larger brain size had better memory, about in proportion. (Remember, the elephant remembers!) Here the demands of language and linguistically coded ideas (as well as every kind of technical and hunting skill) would be a powerful selective force indeed, once spoken language had been incorporated in the behavior as the ideational base.

And the developing views of neurophysiologists (Pribam, 1969; Masland, 1968) seem to give an improved foundation to the idea. Previous beliefs in general saw the cortex as having considerable specificity of function; e.g., roughly speaking, visual stimuli went from retina to geniculate nucleus to visual cortex to association areas, then perhaps to the motor cortex for appropriate response. In the newer view, experiments suggest an earlier distribution of input (e.g., visual or auditory) to various parts of the brain, i.e., association areas, including the frontal lobe, where it is decoded and organized independently of, or before, recognition by the basic visual or auditory

areas. Pribam says the body's muscular re-
sponses to a visual stimulus can be relatively independent of the visual-recognition mechanism, a
finding which is certainly different from the previously accepted wiring system. And Masland shows
that auditory signals go through a straining process
for loudness, pitch, temporal sequence, and the
suppression of white noise, in areas of the brain
other than the specific auditory cortex or the
"speech areas." The importance of all this for
understanding spoken or written language hardly
needs pointing out. And such organization—decoding and recoding—obviously implies "memory," or
conditioning, in the regions involved. Furthermore, these regions are widely distributed in the
cortex and in other parts of the brain (some of
which modulate the level of attention with which
stimuli are experienced). The suggestion by
Pribam is that, in the process of organizing input
information, the nerve impulses entering one of
the organizing regions interact with spontaneous
nerve impulses, or changes in electric potential,
going on in the tissue of the region, to produce a
lasting effect, a "memory," in the molecules
through which the impulses are being transmitted
(see Note 4 on page 162).

This whole modification of views tends to put
"memory" parallel with or ahead of the primary
cortical centers in the reception of input and in its
analysis, so that in fact what arrives at such centers has already been reformulated. "Memory" is
thus more immediately involved in perception and
action instead of being a sort of residual accumulation, however important, in the association areas.
Its role in brain function appears enhanced; in any

case its location is better indicated, as being
widely diffused in the brain, especially those association areas whose function has previously been
less clear. These are the regions, especially the
frontal lobe, which have undergone the most spec-
tacular expansion in men compared to other primates. For natural selection and evolution, we
may look on these areas as having provided a
larger generalized storage, or memory, for the
use of modern man (civilized or not) in his pro-
longed and specialized learning.

The likelihood should be stressed once more
that memory and brain expansion were connected
primarily with language and its role in cultural
and mental processes rather than, as has sometimes been suggested, simply with tool use or
hunting skill. For one thing the component of actual biological evolution, as opposed to cultural,
would have been greater in the progress of linguistic abilities than in those connected with tool-
making. Without at all discounting the degree or
importance of manipulative skill and precision in
hominid development, there is no evidence that it
would have called for dramatic enlargement of
cortical areas in the brain. And memory in other
mammals, especially apes and monkeys, can be
excellent when it is based on visual recognition or
cues of space. The difference between verbally
based and nonverbal memory might be put thus.
When we go to a friend's housewarming we are
given the address, told what route numbers avoid
traffic best, told that Garden Street is the second
past the yellow blinker, and that theirs is the gray
house with a picket fence. This may be a considerable array of symbolized (verbal) information

and is best written down. For later visits, however, we will jettison all this symbolic luggage and
go there just as we, or our dog, go to the post of-
fice over a familiar route which we leave quite un-
verbalized. That is good enough for dogs and
chimpanzees. It is doubtless the gradual, special
elaboration of linguistically based, symbolically
loaded memory which underlies the obvious association between cultural advance and cortical expansion.

Man of today has the only hominid brain we
can observe, either anatomically or in operation.
So we cannot measure the real pace of its development from the first known australopithecine on-
ward. We might ask, without much hope of finding
out, whether there was significant evolution between a late antecedent ("Neanderthal") stage and
that of today. We can note that both Neanderthals
and men of modern type before 35,000 B.C. were
making entirely similar stone tools and, in spite
of some technical distinctions, Mousterian work
grades into that of the Upper Paleolithic in Europe.
Nevertheless, some writers allege a superiority of
modern men over their predecessors. Washburn
(1960) has said that the indirect evidence alone suggests the former were much more intelligent than
Neanderthal man, in spite of the equality in crude
brain size; and Heim (1970), in a detailed study
from skulls, suggests that the Neanderthal brain
was less evolved in critical areas.

Lieberman and his associates (1971, 1972)
have argued with increasing forcefulness that classic Neanderthal man was markedly inferior to ourselves in the capacity for articular speech, being
limited in the range of formant frequencies of the

basic vowels which he was able to produce. Their
conclusion is based on a reconstruction of Neanderthal vocal anatomy using the form of the extended lower jaw and the long, relatively unbent
skull base. The phonological properties of the reconstruction are closer to those of a chimpanzee
or a newborn infant than to adult man, and the argument is persuasive. I find it rather hard to believe, however, partly because other fossil men
seem earlier to have arrived at a skull base more
like modern man's and because the analysis is
based on a single Neanderthal skull; and partly because I would guess (and that is the word) that
primitive language appeared in the australopithecines and, judging from cultural remains, had progressed very far by the late Pleistocene. And, we
have seen, the special form of the whole Neanderthal face may have been more a matter of cold
adaptation than of laggard evolution. But there you
have the argument.

Archaeologists note another difference: a
wealth of ornamental objects and cave art appears
in the Upper Paleolithic of Europe, which is missing from Mousterian levels, and this art even has
cruder parallels in faraway Australia at comparable dates (Gallus, 1968; Pretty, 1971). And
Marshack (1972) has made the striking discovery
of much more subtle expressions of the modern
mind, in records on Upper Paleolithic objects of
bone or stone of what are best understood as observations of phases of the moon.

Taking Stock

For people interested in man's evolution, these
are good times. It is not easy to grasp how much
real advance has been made in the last few decades,
compared to the whole period since Darwin. Major contributions have been the development of absolute dates, without which no reliable ordering of
the rest of the evidence is possible; important
finds of fossils, above all those made by the
Leakey family in East Africa; and rapidly increasing understanding of background knowledge of many
sorts, allowing the emergence of a much broader
and coherent picture altogether.

But human evolution is not like written history
or engineering, with everything connected. We
must realize how far we go in reconstructing the
story, compared to the evidence we actually have.
We use two things: a fossil record which is a tiny
and scattered fraction of the real genealogy; and a
set of models of change and adaptation set up from
what we learn from the anatomy and function of
living primates. We ask the necessary questions

by working out hypotheses as the best explanations
of the material. One man may view such a hypothesis as a tightly reasoned interpretive reconstruction (his own, of course); another may see it as a
farrago of unruminated conjecture. Whole castles
of ideas have tumbled in the past and will again;
the improvement lies in the gradual limitation of
the field of possible error. We deal partly with
fact—and of course there is plenty, in spite of
what I have just said—and largely with probabilities
suggested by the fact; and it is here that different
men with different experience fall out.

Just now, for example, they fall out over when
men and apes—hominids and pongids—first diverged. Ramapithecus is not universally recognized as a hominid. One school (Sarich, 1972) believes that the close biochemical relations of man
and chimpanzee will not allow a separation of
hominids farther back than five million years, if
that—which would make Ramapithecus absolutely
illegitimate as an ancestor.[37] Other hypotheses

---

[37] This is a particularly unreconciled controversy. Sarich, working
also with A. C. Wilson, has amply demonstrated the closeness of man
to the African apes (especially the chimpanzee) in the molecular
makeup of various serum proteins (e.g., globins, albumins), a most
satisfying proof of phylogenetic relationship, which has been referred to earlier. The differences — a few amino acid substitutions in
corresponding molecules — are assumed to be neutral in evolution
and to have occurred by random process since the pongid and hominid lines diverged. Translating this into absolute time brings the controversy. Sarich assumes the rate of replacement to be regular, and
calibrates it by earlier events, such as the split between hominoid and
Old World monkeys, perhaps 37 million years ago, and by other
checks. Others have objected to computing rates this way. For example, Lovejoy et al. (1972) believe that the longer generation length
of apes and men should be considered, slowing the rate in later times,
and providing an estimate for the ape-man separation of about 14
million years rather than five million.

relating to behavior, or geology, or locomotion
(e.g., knuckle-walking) may favor a recent separation and thus they, and their supporters, find mu-
tual accord and so tend to form a "school." On the
other hand, paleontologists satisfied with the hominid status of Ramapithecus cannot even negotiate
with this school because of the unquestioned dates
of over 10 million years for the fossils; instead
they find agreement, for example, with those who
would be happier with a longer history, preceding
the australopiths, in which the lower body changed
structure for bipedalism.

In the latest period we know that the Mousterian cultures preceded those of the Upper Paleolithic because that is how they lie in the ground.
We also know the skull shapes of Neanderthal and
Upper Paleolithic men. But students disagree
about the precise relations of the two populations.
We have no good way of saying something like "Neanderthal man was 40 percent different from modern man" and then going on to show it in such a
way that others are obliged to accept it as a fact.

So it goes, though with gradual establishment
of pieces of solid ground. Even these may be
shaken. While there continues to be plenty of disagreement about lines of descent going through the
australopith stage, it had been felt that Homo, and
larger brains, waited until perhaps a million years
ago to make his appearance. Now Richard Leakey
has produced an unexpectedly large brain at more
than 2.6 million years, and his coworker, Kay
Behrensmeyer, has tools from that period of time—
another surprise.

New dates for longer-known specimens also
cause minor quakes, in view of so much ignorance

of absolute ages. The possible earlier dating of
the Rhodesian man and the Petralona skull are important examples, and we shall certainly do better
in the future. Primarily, however, progress depends on new finds of fossils. Today, contexts of
dating, cultural material, and general environment can be determined so much more fully, compared to the time when most of the skulls came to
light, that each new discovery carries a much
greater freight of information. Unfortunately we
cannot force the fossils to appear. We can only
press the search in obviously good situations, as
the Leakeys have been doing in Africa, and as Indonesian scientists are doing around the Sangiran
Dome in Java.

Nevertheless discoveries do come more rapidly. And work leading to better understanding
of real differences among the known fossil men,
and trying to see their real meaning for populations present in a given place and time, does con-
stitute advance. So, while we wait for new finds
to fill out all those blank spaces on the charts,
what can best be done is fresh study of older material, and fresh work in the field to pick up the
signs of date and of ecology which the earlier
workers, who found the actual fossils, missed or
ignored. What was done archaeologically to give
Broken Hill man a cultural home, long after his
skull first came to light, is an example. Musgrave's restudy of the Neanderthal hand is another.
Studies of function and anatomy, in man and primates, is also important. There is a great deal
of such work still to be done, indoors and out—
more than is easily realized.

Notes

1. It is hardly possible at present to use "Middle
Pleistocene" (= post-Villafranchian generally) as
representing agreement among scholars, or with
precision either in specifying time zones within different parts of the Old World or in relating these
parts chronologically to one another. These divisions of the Pleistocene refer, of course, to faunal
divisions (essentially fossil mammals); and quite
different kinds of genera or of species will have
been present, or will mark major changes, in different regions.

The problem comes to the surface in considering Java. There is agreement as to the Middle
Pleistocene character of the Trinil fauna (of the
Kabuh beds). Von Koenigswald has long maintained
that the Djetis fauna, in the Putjangan beds below,
is Lower Pleistocene. However, he himself long
ago made the distinction between the Siva-Malayan
fauna, derived from India and earlier than all the
above in Java, and the Sino-Malayan fauna, stem-
ming from a later, different migration from China

and arriving in Java as the Djetis fauna. Hooijer,
analyzing the Djetis animals in much detail and in
several studies, has never doubted that they are
post-Villafranchian, i.e., Middle Pleistocene, in
character. And the Djetis and Trinil faunas are
not profoundly different, so the correlation with
the geological divisions (Putjangan and Kabuh) is
probably only general. If a date of 1.9 million
years in the Putjangan beds actually marked a
Middle Pleistocene fauna, this would obviously be
an earlier Middle Pleistocene than that at Olduvai,
and far older than that in Europe.

2.   Contrast between supposedly early and later
Java specimens might be overdone by error, since
a strict time ordering of the fossils is not possible.
Many have been found by villagers, and thus never
traced to the exact finding spot and so precisely
located as to level, but only judged by general locality or evidence. Even at the original site,
where no Djetis fossils are known, the silts of the
Solo River at Trinil may be redeposited, thus containing materials of different ages. Thus the date
of the first skull is not a positive association
(though it is not likely to belong to anything but the
Trinil zone originally), nor—and this is important—
is there a positive association for the leg bone
found with it. Pithecanthropus III, an apparent
female and entirely like I and II in conformation,
was assigned to the Trinil zone on finding, after
which von Koenigswald lost much of his data in
World War II. He now thinks (von Koenigswald and
Ghosh, 1973) that material with which the skull
cap was filled shows that it belonged to Notopuro
beds just above the Kabuh (Trinil) series, and
would thus be a late specimen. On the other hand,

recent tests of the skull for relative amounts of
nitrogen, fluorine, etc., by Kenneth Oakley and
Theya Molleson (for which information I thank
them) agree best with results on other fossils from
the Djetis, which suggests that the skull might be
early.

3.   A previously proposed "frostbite" or "cold-
engineered" facial model (Coon, Garn, and Birdsell, 1950), the broad, flat, fat-padded face of
arctic Mongoloids and Eskimos, is controverted by
Steegmann's physiological evidence. This second
model, he supposes, is primarily suited to the
protection of internal structures including the eyes,
respiratory passages, and brain arteries. One
may suppose that this form might be selected for
by another kind of cold attack: extreme, penetrating dry cold with a high wind-chill factor. In such
a case, two lines of selective response might have
been followed, perhaps in two populations differing
facially to begin with. Steegmann suggests that the
"internal structures" model might be more suitable
in the case of communities already able to protect
themselves by head and face gear against ordinary
exposure of these parts to less extreme low tem-
peratures. At any rate, he considers that the Neanderthal face fits both models at the same time
fairly well; apparently he is taking note of the
large nose and large maxillary sinuses, as well
as of facial protrusion and slanting cheekbones.

4.   A simile used by this school of interpretation
of memory is the hologram. In the familiar record of the image of an object (a photographic negative or its model, the retina of the eye), light reflected from an object is of varied wavelengths,

and the image is focused on the negative by a lens, point for point, so that a little bit of the negative holds only a little bit of the image or picture. In a hologram the wavelength (color) of the light from an object is highly controlled, as in a laser, and a second, reference, source of the same light is directed to the negative, which therefore is able to record the image as the interference pattern of light from the two sources. This "image" is not recognizable visually. But the proper image can be reconstructed by use of the reference light; and because all of the image is dispersed over all of the hologram, a small part of the latter will reproduce a less clear but complete version of the whole image. (And more than one image can be stored in one hologram by using optical controls.) Similitudes for the brain suggested by neurophysiologists are: (a) the passage of a "wave front" of nervous impulses (from the optic or auditory nerves, for example) across synapses between cortical cells, hypothetically interacting with standing wave impulses to produce lasting molecular configurations, and (b) the distribution of such recording widely in the brain so that a coherent "memory" can be evoked from a small part of the brain. The ability of "memories" to survive in brains which have suffered destruction of large and varied parts of the cortex was one fact which focused attention on this kind of interpretation.

References

Andrews, P. (1971). "Ramapithecus wickeri mandible from Fort Ternan, Kenya." Nature, 231: 192-194.

Bada, J. L., and R. Protsch (1973). "Racemization reaction of aspartic acid and its use in dating fossil bones." Proceedings, National Academy of Sciences, 70: 1331-1334.

Bilsborough, A. (1971). "Evolutionary change in the hominoid maxilla." Man, 6: 473-485.

Birdsell, J. B. (1967). "Preliminary data on the trihybrid origin of the Australian aborigines." Archaeology and Physical Anthropology in Oceania, 2: 100-155.

Bowler, J. M., A. G. Thorne, and H. A. Polach (1972). "Pleistocene man in Australia: age and significance of the Mungo skeleton." Nature, 240: 48-50.

Brace, C. L. (1964). "The fact of the 'classic' Neanderthals: a consideration of hominid catastrophism." Current Anthropology, 5: 3-43.

Brace, C. L., and P. E. Mahler (1971). "Post-Pleistocene changes in the human dentition." American Journal of Physical Anthropology, 34: 191-204.

Breitinger, E. (1955). "Das Schädelfragment von Swanscombe und das 'Prasesapiensproblem'. Mit einer Rekonstruktion des Calvariums und Bemerkungen zur stammesgeschichtlichen Stellung des Fundes." Mitteilungen der Anthropologischer Gesellschaft in Wien, 84-85: 1-45.

Brose, D. S., and M. H. Wolpoff (1971). "Early Upper Paleolithic man and late Middle Paleolithic tools." American Anthropologist, 73: 1156-1194.

Butzer, K. W., F. H. Brown, and D. L. Thurber (1969). "Horizontal sediments of the lower Omo valley: the Kibish formation." Quarternaria, 11: 15-29.

Bygott, J. D. (1972). "Cannibalism among wild chimpanzees." Nature, 238: 410-411.

Campbell, B. G. (1963). "Quantitative taxonomy and human evolution." In S. L. Washburn, ed., Classification and Human Evolution, pp. 50-74. Chicago: Aldine.

Clark, J. D. (1959). "Further excavations at Broken Hill, Northern Rhodesia." Journal of the Royal Anthropological Institute, 89: 201-232.

Clark, W. E. Le Gros, and L. S. B. Leakey (1951). The Miocene Hominoidea of East Africa. Fossil Mammals of Africa, No. 1. British Museum of Natural History.

Clarke, R. J., F. C. Howell, and C. K. Brain (1970). "More evidence of an advanced hominid at Swartkrans." Nature, 225: 1219-1222.

Conroy, G. C. (1972). "Problems in the interpretation of Ramapithecus: with special reference to anterior tooth reduction." American Journal of Physical Anthropology, 37: 41-48.

Conroy, G. C., and J. G. Fleagle (1972). "Locomotor behaviour in living and fossil pongids." Nature, 237: 103-104.

Cooke, H. B. S. (1972). "Pleistocene chronology: long or short?" Maritime Sediments, Vol. 8, No. 1, pp. 1-12.

Cooke, H. B. S., and V. J. Maglio (1972). "Plio-Pleistocene stratigraphy in East Africa in relation to proboscidean and suid evolution." In W. W. Bishop and J. A. Miller, eds., Calibration of Hominoid Evolution, pp. 303-329. Scottish Academic Press.

Coon, C. S. (1939). The Races of Europe. New York: Macmillan.

Coon, C. S. (1962). The Origin of Races. New York: Knopf.

Coon, C. S., S. M. Garn, and J. B. Birdsell (1950). Races. A Study of the Problems of Race Formation in Man. Springfield, Ill.: Thomas.

Dart, R. A. (1949). "The predatory implemental technique of Australopithecus." American Journal of Physical Anthropology, 7: 1-38.

Davis, P. R. (1964). "Hominid fossils from Bed I, Olduvai Gorge, Tanganyika." Nature, 201: 967-968.

Day, M. H. (1969). "Femoral fragment of a robust australopithecine from Olduvai Gorge, Tanzania." Nature, 221: 230-233.

Day, M. H. (1971). "Postcranial remains of Homo erectus from Bed IV, Olduvai Gorge, Tanzania." Nature, 232: 383-387.

Day, M. H. (1972). "The Omo human skeletal remains." In F. Bordes, ed., The Origin of Homo Sapiens, pp. 31-35. Proceedings, Paris Symposium, 2-5 September, 1969, organized by UNESCO with INQUA.

Day, M. H., and J. R. Napier (1966). "A hominid toe bone from Bed I, Olduvai Gorge, Tanzania." Nature, 211: 929-930.

Day, M. H., and B. A. Wood (1968). "Functional affinities of the Olduvai hominid 8 talus." Man, 3: 440-455.

Day, M. H., and B. A. Wood (1969). "Hominoid tali from East Africa." Nature, 222: 591-592.

DeVore, I., and S. L. Washburn (1963). "Baboon ecology and human evolution." In F. C. Howell and F. Bourlière, eds., African Ecology and Human Evolution, pp. 335-367. Chicago: Aldine.

Dubois, E. (1894). Pithecanthropus erectus, eine menschenaehnliche Uebergangsform aus Java. Batavia.

Elftman, H. (1944). "The bipedal walking of the chimpanzee." Journal of Mammalogy, 25: 67-71.

Elftman, H., and J. Manter (1935a). "Chimpanzee and human feet in bipedal walking." American Journal of Physical Anthropology, 20: 69-79.

Elftman, H., and J. Manter (1935b). "The evolution of the human foot with especial reference to the joints." Journal of Anatomy, 70: 56-67.

Ennouchi, E. (1968). "Le deuxième crâne de l'homme d'Irhoud." Annales de Paléontologie, 54: 117-128.

Ennouchi, E. (1969). "Présence d'un enfant Neanderthalien au Jebel Irhoud (Maroc)." Annales de Paléontologie, 55: 251-265.

Gallus, A. (1968). "Parietal art in Koonalda Cave, Nullabor Plain, South Australia." Helictite, 6: 43-49.

Gardner, R. A., and B. T. Gardner (1969). "Teaching sign language to a chimpanzee." Science, 165: 664-672.

Geschwind, N. (1972). "Language and the brain." Scientific American, April, 1972, pp. 76-83.

Goodall, J. (1963). "Feeding behaviour of wild chimpanzees." The Primates. Symposia of the Zoological Society of London, No. 10, pp. 39-47.

Heim, J. -L. (1970). "L'encéphale néandertalien de l'homme de La Ferrassie." L'Anthropologie, 74(7-8): 527-572.

Hemmer, H. (1972). "Notes sur la position phylétique de l'homme de Petralona." L'Anthropologie, 76: 155-162.

Hewes, G. W. (1961). "Food transport and the origin of human bipedalism." American Anthropologist, 63: 687-710.

Higgs, E. S. (1961). "Some Pleistocene faunas of the Mediterranean coastal areas." Proceedings of the Prehistoric Society, 27: 144-154.

Hill, J. H. (1972). "On the evolutionary foundations of language." American Anthropologist, 74: 308-317.

Hockett, C. F. (1960). "The origin of speech." Scientific American, 203, No. 3, pp. 88-96.

Holloway, R. L., Jr. (1966). "Cranial capacity, neural reorganization, and hominid evolution: a search for more suitable parameters." American Anthropologist, 68: 103-121.

Holloway, R. L., Jr. (1972). "Australopithecine enodcasts, brain evolution in the Hominoidea, and a model of hominid evolution." In R. Tuttle, ed., The Functional and Evolutionary Biology of Primates, pp. 185-203. Chicago: Aldine.

Holloway, R. L., Jr. (1973). "New endocranial values for the East African early hominids." Nature, 243: 97-99.

Howell, F. C. (1952). "Pleistocene glacial ecology and the evolution of 'Classic Neanderthal' man." Southwestern Journal of Anthropology, 8: 377-410.

Howell, F. C., G. H. Cole, M. R. Kleindienst, B. J. Szabo, and K. P. Oakley (1972). "Uranium-series dating of bone from the Isimila prehistoric site, Tanzania." Nature, 237: 51-52.

Howells, W. W. (1970). "Mount Carmel Man: morphological relationships." Proceedings, 8th International Congress of Anthropological and

Ethnological Sciences. Tokyo and Kyoto, 1968.
Vol. 1, pp. 269-272.

Howells, W. W. (1973a). Cranial Variation in
Man. A Study by Multivariate Analysis. Peabody
Museum Papers, Vol. 67.

Howells, W. W. (1973b). The Pacific Islanders.
London: Weidenfeld and Nicolson.

Hrdlička, A. (1927). "The Neanderthal phase of
man." Journal of the Royal Anthropological Institute, 57: 249-274.

Isaac, G. L. (1972). "Chronology and the tempo
of cultural change during the Pleistocene." In
W. W. Bishop and J. A. Miller, Calibration of
Hominoid Evolution, pp. 381-430. Scottish Academic Press.

Isaac, G. L., R. E. F. Leakey, and A. K.
Behrensmeyer (1971). "Archeological traces of
early hominid activities, east of Lake Rudolf,
Kenya." Science, 173: 1129-1134.

Jacob, T., and G. H. Curtis (1971). "Preliminary potassium-argon dating of early man in Java."
Contributions of the University of California Archaeological Research Facility, No. 12, p. 50.

Jantschke, F. (1972). Orang-utans in zoolog-
ischen Garten. Munich: Piper.

Jenkins, F. A., Jr. (1972). "Chimpanzee bipedalism: cineradiographic analysis and implications
for the evolution of gait." Science, 178: 877-879.

Jolly, C. J. (1970). "The seed-eaters: a new
model of hominid behavioral differentiation based
on a baboon analogy." Man, 5: 5-26.

Kipp, F. A. (1955). "Die Entstehung der menschlichen Lautbildungsfähigkeit als Evolutionsproblem." Experientia, 11: 89-94.

Klein, R. B. (in press). Ice Age Hunters of the Ukraine. University of Chicago Press.

Klein, R. G. (1970). "Problems in the study of the Middle Stone Age of South Africa." South African Archaeological Bulletin, XXV: 127-135.

Klein, R. G. (1974, in press). "Environment and subsistence of prehistoric man in the southern Cape Province, South Africa." World Archaeology, Vol. 5.

Klein, R. G. (in press). On the geological antiquity of Rhodesian man. Nature.

von Koenigswald, G. H. R. (1967). "Neue Dokumente zur menschlichen Stammesgeschichte." Bericht der Schweizerischen Palaeontologischen Gesellschaft, 46. Jahresversammlung. Ecologae geologicae Helvetiae, 60: 641-655.

von Koenigswald, G. H. R., and A. K. Ghosh (1973). "Stone implements from the Trinil beds of Snagiran, Central Java." Proceedings, Koninklijke Akademie van Wetenschappen, series B, 76: 1-34.

Kortlandt, A., and J. C. J. van Zon (1969). "The present state of research on the dehumanization hypothesis of African ape evolution." Proceedings, 2nd International Congress of Primatology, Vol. 3, pp. 10-13.

Kurtén, B. (1972). The Age of Mammals. New York: Columbia University Press.

Kurth, G. (1967). "Implications of primate pale-ontology for behavior." In J. Spuhler, ed., Genetic Diversity and Human Behavior, pp. 199-213. Chicago: Aldine.

Leakey, L. S. B., P. V. Tobias, and J. R. Napier (1964). "A new species of the genus Homo from Olduvai Gorge." Nature, 202: 7-9.

Leakey, M. D. (1970a). "Stone artefacts from Swartkrans." Nature, 225: 1222-1225.

Leakey, M. D. (1970b). "Early artefacts from the Koobi Fora area." Nature, 226: 228-230.

Leakey, M. D. (1971). "Discovery of postcranial remains of Homo erectus and associated artefacts in Bed IV at Olduvai Gorge, Tanzania." Nature, 232: 380-383.

Leakey, M. D., R. J. Clarke, and L. S. B. Leakey (1971). "New hominid skull from Bed I, Olduvai Gorge, Tanzania." Nature, 232: 308-312.

Leakey, R. E. F. (1971). "Further evidence of Lower Pleistocene hominids from East Rudolf, North Kenya." Nature, 231: 241-245.

Lewis, O. J. (1972). "Evolution of the hominoid wrist." In R. Tuttle, ed., The Functional and Evolutionary Biology of Primates, pp. 207-222. Chicago: Aldine-Atherton.

Lieberman, P. (1968). "Primate vocalizations and human linguistic ability." Journal of Acoustical Society of America, 44: 1574-1584. Reprinted in S. L. Washburn and P. Dolhinow, Perspectives in Human Evolution, 2: 444-468.

Lieberman, P., and E. S. Crelin (1971). "On the speech of Neanderthal man." Linguistic Inquiry, 2: 203-222.

Lieberman, P., E. S. Crelin, and D. H. Klatt (1972). "Phonetic ability and related anatomy of the newborn and adult human, Neanderthal man, and the chimpanzee." American Anthropologist, 74: 287-317.

Lovejoy, C. O., A. H. Burstein, and K. G. Heiple (1972). "Primate phylogeny and immunological distance." Science, 176: 803-805.

de Lumley, H., and M. -A. de Lumley (1971). "Découverte de restes humains anténéandertaliens datés du début du Riss à la Caune de l'Arago (Tautavel, Pyrénées-Orientales)." Comptes Rendus de l'Académie des Sciences de Paris, 272: 1739-1742.

Maier, W. (1972). "The first complete skull of Simopithecus darti from Makapansgat, South Africa, and its systematic position." Journal of Human Evolution, 1: 395-405.

Marshack, A. (1972). The Roots of Civilization. New York: McGraw-Hill.

Martin, R. (1928). Lehrbuch der Anthropologie, 2nd ed. 3 Vols. Jena: G. Fischer.

Masland, R. L. (1968). "Some neurological processes underlying language." Annals of Otology, Rhinology and Laryngology, 27: 787-804. Reprinted in S. L. Washburn and P. Dolinhow, eds., Perspectives in Human Evolution, 2: 421-437.

McCown, T. D., and A. Keith (1939). The Stone Age of Mount Carmel. The Fossil Human Remains from the Levallois-Mousterian, Vol. II. Oxford: Clarendon.

McHenry, H. M. (1972). The Postcranial Skeleton of Early Pleistocene Hominids. Ph.D. thesis, Harvard University.

McHenry, H. M. (1973). "Early hominid humerus from East Rudolf, Kenya." Science, 180: 739-741.

McNeish, R. (1971). "Early man in the Andes." Scientific American, April, 1971, pp. 36-46.

Merrick, H. V., J. de Heinzelin, P. Haesaerts, and F. C. Howell (1973). "Archaeological occurrences of early Pleistocene age from the Shungura formation, lower Omo valley, Ethiopia." Nature, 242: 572-575.

Morton, D. J. (1935). The Human Foot. Its Evolution, Physiology and Functional Disorders. New York: Columbia University Press.

Movius, H. L., Jr. (1972). "Radiocarbon dating of the Upper Paleolithic sequence at the Abri Pataud, Les Eyzies (Dordogne)." In F. Bordes, ed., The Origin of Homo sapiens, pp. 253-260. Proceedings of the Paris Symposium, 2 - 5 September, 1970, organized by UNESCO and INQUA. Paris: UNESCO.

Musgrave, J. (1971). "How dextrous was Neanderthal man?" Nature, 233: 538-541.

Napier, J. (1962). "The evolution of the hand." Scientific American, 207: 56-62. Reprinted in W. S. Laughlin and R. H. Osborne, Human Variation and Origins. San Francisco: Freeman.

Napier, J. R. (1964). "The evolution of bipedal walking in the hominids." Archives de Biologie, 75: 673-708.

Napier, J. (1967). "The antiquity of human walking." Scientific American, 216: 56-66. Reprinted in W. S. Laughlin and R. H. Osborne, Human Variation and Origins. San Francisco: Freeman.

Nissen, H. W. (1931). "A field study of chimpanzee." Comparative Psychology Monographs, Vol. 8, pp. 1-122.

Oakley, K. P. (1964). Frameworks for Dating Fossil Man. London: Weidenfeld and Nicolson.

Oakley, K. P., and B. G. Campbell (1967). Catalogue of Fossil Hominids. Part I: Africa. London: British Museum of Natural History.

Oxnard, C. E. (1972). "Some African fossil foot bones: a note on the interpolation of fossils into a matrix of extant species." American Journal of Physical Anthropology, 37: 3-12.

Patterson, B., and W. W. Howells (1967). "Hominid humeral fragment from early Pleistocene of northwestern Kenya." Science, 156: 64-66.

Piveteau, J. (1967). "Un parietal humain de la grotte du Lazaret (Alpes-Maritimes)." Annales de Paléontolgie, 53: 165-199.

Piveteau, J. (1970). Les Grottes de La Chaise (Charente). Paléontologie humaine. I. L'homme de l'Abri Suard. Annales de Paléontologie (Vertébrés), 56: 174-225.

Premack, A. J., and D. Premack (1972).
"Teaching language to an ape." Scientific American, October, 1972, pp. 92-99.

Premack, D. (1971). "Language in chimpanzee?" Science, 172: 808-822.

Preuschoft, H. (1971). "Body posture and mode of locomotion in early Pleistocene hominids." Folia Primatologia, 14: 209-240.

Preuschoft, H. (1973). "Body posture and locomotion in some East African Miocene Dryopithecinae. In M. H. Day, ed., Human Evolution, pp. 13-46. Symposia of the Society for the Study of Human Biology, Vol. 11.

Pribam, K. H. (1969). "The neurophysiology of remembering." Scientific American, 220: 73-86.

Protsch, R. R. R. (1973). "The dating of Upper Pleistocene Subsaharan fossil hominids and their place in human evolution: the morphological and archaeological implications." Ph.D. thesis, University of California, Los Angeles.

Rightmire, G. P. (1972). "Multivariate analysis of an early hominid metacarpal from Swartkrans." Science, 176: 159-161.

Robinson, B. W. (1972). "Anatomical and physiological contrasts between human and other primate vocalizations." In S. L. Washburn and P. Dolhinow, Perspectives on Human Evolution, 2: 438-443.

Robinson, J. T. (1953a). "Meganthropus, australopithecines and hominids." American Journal of Physical Anthropology, 11: 1-38.

Robinson, J. T. (1953b). "Telanthropus and its phylogenetic significance." American Journal of Physical Anthropology, 11: 445-501.

Robinson, J. T. (1956). "The dentition of the Australopithecinae." Memoirs, Transvaal Museum Pretoria, No. 9: 1-179.

Robinson, J. T., L. Freedman, and B. A. Sigmon (1972). "Some aspects of pongid and hominid bipedality." Journal of Human Evolution, 1: 361-369.

Sarich, V. M. (1972). "A molecular approach to the question of human origins." In P. Dolhinow and V. M. Sarich, eds., Background for Man, pp. 60-81. Boston: Little Brown.

Sartono, S. (1971). "Observations on a new skull of Pithecanthropus erectus (Pithecanthropus VIII) from Sangiran, central Java." Proceedings, Koninkl. Nederl. Akad. van Wetenschappen, Series b74, No. 2, pp. 185-194.

Sigmon, B. A. (1971). "Bipedal behavior and the emergence of erect posture in man." American Journal of Physical Anthropology, 34: 55-60.

Simons, E. L. (1969). "Late Miocene hominid from Fort Ternan, Kenya." Nature, 221: 448-451.

Simons, E. L., and D. R. Pilbeam (1965). "Preliminary revision of the Dryopithecinae (Pongidae, Anthropoidea)." Folia Primatologia, 3: 81-152.

Simons, E. L., and D. R. Pilbeam (1972). "Hominoid paleoprimatology." In R. Tuttle, The Functional and Evolutionary Biology of Primates, pp. 36-62. Chicago: Aldine-Atherton.

Simons, E. L., and I. Tattersall (1971).
"Origin of the family of man." Ventures (Magazine of the Yale Graduate School), XI: 47-55.

Steegmann, A. T., Jr. (1972). "Cold response, body form, and craniofacial shape in two racial groups of Hawaii." American Journal of Physical Anthropology, 37: 193-221.

Stewart, T. D. (1960). "Form of the pubic bone in Neanderthal man." Science, 131: 1437-1438.

Stewart, T. D. (1961). "A neglected primitive feature of the Swanscombe skull." In Homenaje a Pablo Martinez del Rio, pp. 207-217. Reprinted in C. E. Ovey, ed., The Swanscombe Skull. Occasional Papers of the Royal Anthropology Institute, No. 20, pp. 151-160.

Stewart, T. D. (1962). "Neanderthal scapulae with special attention to the Shanidar Neanderthals from Iraq." Anthropos, 57: 781-800.

Stewart, T. D. (1964). "The scapula of the first recognized Neanderthal skeleton." Bonner Jahrbuch, 164: 1-14.

Stewart. T. D. (1970). "The evolution of man in Asia as seen in the lower jaw." Proceedings, 8th International Congress of Anthropological and Ethnological Sciences, Vol. 1, pp. 263-266.

Suzuki, H. (1970). "The Amud Man and the Shanidar Man." Proceedings, 8th International Congress of Anthropological and Ethnological Sciences, Vol. 1, pp. 273-278.

Taylor, C. R., and V. J. Rowntree (1973). "Running on two legs: which consumes more energy?" Science, 179: 186-187.

Teleki, G. (1973). "The omnivorous chimpanzee." Scientific American, January, 228: 32-42.

Thoma, A. (1964). "Die Entstehung der Mongoliden." Homo, 15: 1-22.

Thoma, A. (1966). "L'occipital de l'homme mindélien de Vértesszöllös." L'Anthropologie, 70/5-6: 495-534.

Thoma, A. (1969). "Biometrische Studie über das occipitale von Vértesszöllös." Zeitschrift für Morphologie und Anthropologie, 60: 229-241.

Thoma, A. (1972). "On Vértesszöllös man." Nature, 236: 464-465.

Thorne, A. G., and P. G. Macumber (1972). "Discoveries of late Pleistocene man at Kow Swamp, Australia." Nature, 238: 316-319.

Thurber, D. L. (1972). "Problems of dating non-woody material from continental environments." In W. W. Bishop and J. A. Miller, Calibration of Hominoid Evolution, pp. 1-17. Edinburgh: Scottish Academic Press.

Tobias, P. V. (1966). "The distinctiveness of Homo habilis." Nature, 209: 953-957.

Tobias, P. V. (1967). The cranium and maxillary dentition of Australopithecus (Zinjanthropus) boisei. Olduvai Gorge, Vol. 2. Cambridge: University Press.

Tobias, P. V. (1971a). The Brain in Hominid Evolution. New York: Columbia University Press.

Tobias, P. V. (1971b). "The distribution of cranial capacity values among living hominoids."

Proceedings 3rd International Congress of Primatology, Zurich, 1970, Vol. 1: 18-35.

Tuttle, R. H. (1969). "Knuckle-walking and the problem of human origins." Science, 166: 953-961.

Vallois, H.-V. (1932). "L'omoplate humaine." Bulletins et Mémoires de la Société d'Anthropologie de Paris, Vol. 3, Series 8, Pts. 1-3: 3-153.

Vallois, H.-V. (1969). "Le temporal néandertalien H 27 de La Quina. Étude anthropologique." L'Anthropologie, 73: 365-400, 525-544.

Vallois, H.-V., and B. Vandermeersch (1972). "Le crane moustérien de Qafzeh (Homo VI). Étude anthropologique." L'Anthropologie, 76: 71-96.

Vandermeersch, B. (1972). "Récentes découvertes de squelettes humains à Qafzeh (Israel): essai d'interpretation." In F. Bordes, ed., The Origin of Homo sapiens, pp. 49-54. Proceedings of Paris Symposium, 2 - 5 September, 1970, organized by UNESCO and INQUA. Paris: UNESCO.

Vlček, E. (1967). "Die Sinus frontales bei europäischen Neanderthalern." Anthropologischer Anzeiger, 30/2-3: 166-189.

Vogel, J. C., and P. B. Beaumont (1972). "Revised radiocarbon chronology for the Stone Age in South Africa." Nature, 237: 50-51.

Washburn, S. L. (1960). "Tools and human evolution." Scientific American, 203/3: 62-87.

Washburn, S. L. (1968). The Study of Human Evolution. Condon Lectures, University of Oregon.

Washburn, S. L., and C. S. Lancaster (1968).
"The evolution of hunting." In S. L. Washburn and
P. C. Jay, Perspectives on Human Evolution, Vol.
1, pp. 213-229. New York: Holt, Rinehart and
Winston.

Weidenreich, F. (1928). "Entwicklungs- under
Rassetypen des Homo primigenius." Natur und
Museum, 58: 1-13, 51-62.

Weidenreich, F. (1941). "The extremity bones of
Sinanthropus pekinensis." Paleontologica Sinica,
New Series D., No. 5, Whole Series No. 116.

Weidenreich, F. (1946). Apes, Giants and Man.
Chicago: University of Chicago.

Weiner, J. S. (1958). "The pattern of evolutionary development of the genus Homo." South African Journal of Medical Science, 23: 111-120.

Weiner, J. S., and B. G. Campbell (1964). "The
taxonomic status of the Swanscombe skull." In
C. D. Overy, ed., The Swanscombe skull. A sur-
vey of research on a Pleistocene site. Royal Anthropological Institute, Occasional Paper No. 20,
pp. 175-209.

Wolpoff, M. H. (1971). "Vértesszöllös and the
presapiens theory." American Journal of Physical Anthropology, 35: 209-216.

Wright, R. V. W. (1972). "Imitative learning of
a flaked stone technology—the case of an orangu-
tan." Mankind, 8: 296-306.

Index